No One May Ever Have the Same Knowledge Again

COURTESY MOUNT WILSON OBSERVATORY

FRONTISPIECE

The 100-inch Mirror of the Hooker Telescope

CONTRIBUTIONS FROM THE
MUSEUM OF JURASSIC TECHNOLOGY

NO ONE MAY EVER HAVE THE SAME KNOWLEDGE AGAIN

Letters to Mount Wilson Observatory
1915-1935

PUBLISHED BY THE TRUSTEES

Funded in part by the City of Los Angeles, Cultural Affairs Department

Society For the Diffusion of Useful Information Press
9091 Divide Place West Covina, California OX2 6DP

Billings Bogata Bhopal Beirut
Bowling Green Buenos Aires Campton
Dayton Dar es Salaam Düsseldorf
Fort Wayne Indianapolis Lincoln
Mal en Beg Mal en Mor
Nannin Pretoria Teheran
Socorro Terra Haute Ulster

Published in the United States by
The Society for the Diffusion of Useful Information

Published in cooperation with the Visitors to the Museum
by the Delegates of the Press

British Library Cataloguing-in-Publication Data
I. Drake, Diana.
II. Simons, Sarah.
III. Rossi, M. Francis.
IV. Wilson, David.
069' .06937'75 AP 101.0794

Library of Congress Cataloging-in-Publication Data
Main entry under title:
1. No One May Ever Have the Same Knowldege Again
Letters to Mount Wilson Observatory
Includes plates.
1. Museum of Jurassic Technology. 2. Simons, Sarah 1959 –
3. Rossi, M. Francis 1967– 4. Drake, Diana 1947–
5. Wilson, David 1946–
II. Museum of Jurassic Technology
AP 101.0794 093' .06937'7 88-73931
Set in Ventura Publisher.
Printed by McNaughton & Gunn in the
United States of America

CONTENTS

FOREWORD

The material for this Leaflet is derived from an exhibit bearing the name *No One May Ever Have the Same Knowledge Again - Letters to Mount Wilson Observatory, 1915-1935* housed at The Museum of Jurassic Technology in Los Angeles, California. The exhibit places in public view some twenty two of the letters contained in this catalogue.

The Museum wishes to thank Mount Wilson Observatory and the Carnegie Institution for making the letters available for this exhibition. In addition we would like to thank Tony Misch and Larry Webster for bringing the letters to our attention nearly a decade ago and providing many hours of support on our repeated trips to the observatory. Finally, we wish to acknowledge the extraordinary efforts of Sarah Simons whose uncanny ability for deciphering often impenetrable handwriting and whose tireless work made the exhibition and this catalogue possible.

INTRODUCTION

The Mount Wilson Observatory lies a short distance north east of Pasadena California at an altitude of 5,704 feet above the sea in the range of mountains known as the Sierra Madre. The idea for the observatory was conceived in the early years of the twentieth century by Dr. George Ellery Hale, a uniquely brilliant and visionary astronomer then Director of the Yerkes Observatory in Racine, Wisconsin.

Frustrated by the limitations of solar observation provided by the often overcast and perpetually hazy Wisconsin skies, Dr. Hale was intent on establishing an observatory at a location with as nearly ideal climatic and atmospheric conditions as could be had within reasonable distance of a city where instruments could be made and supplies secured.

In his search for such a site, Dr. Hale approached a colleague, W. J. Hussey of the Lick Observatory on Mount Hamilton near San Jose, California, for recommendations. Professor Hussey had in the preceeding years researched a number of locations from Australia to Arizona for observation sites and, based on his search, Professor Hussey recommended the Sierra Madre and Mount Wilson in particular to Dr. Hale for a site for his proposed solar observatory. In the winter of 1903 - 04 with the aid of a 3 1/4 inch refractor, Dr. Hale thoroughly tested the site to determine the suitability of the location for astronomical observations. The following year the Snow solar telescope from the Yerkes Observatory was installed on the Mountain.

Dr. Hale soon realized the stellar as well as solar observation potential of the site and by 1908, through the unflagging energies of Dr. Hale as well as the generosity of the Carnegie Institution, a 60 inch reflector, the largest actively used telescope in existence, was routinely probing the interstellar depths of the California night sky.

Beginning in 1905 the Observatory regularly published the results of its research through a series of papers in a number of scientific journals including *The Astrophysical Journal*, *Publications of the Astronomical Society of the Pacific* and *The Proceedings of the National Academy of Sciences*. Taken together, these papers constitute the massive *Contributions from the Mount Wilson Observatory* - a venerable collection of information which contains a large percentage of the major astrophysical discoveries of the years of the first half of this century.

Almost immediately certain of the observatory's findings began to trickle down to the lay public through the popular press.

The extraordinary mechanical methods used with the large instruments as well as the concrete results obtained by the astronomers began to be of general interest. Through the teens and especially after the completion of the extraordinary 100 inch telescope in 1918, the trickle of information approached a stream. And through the 1920's and into the 30's, fueled by the astonishing discoveries made by Hale, Hubble, Michelson, and their contemporaries, the flow of public interest became a torrent. By the beginning of the third decade of this century some 20,000 people annually visited the observatory and tens of thousands of others followed the astronomers' progress from afar.

As early as 1911, the astronomers at Mount Wilson began receiving letters from people all around the world, people from all walks of life, educated as well as uneducated. Many of the letters were simple expressions of appreciation and awe for the work that the astronomers were accomplishing. There was, however, another class of letter. These letters were communications to the astronomers by individuals who felt, often with a great degree of earnestness, that they were in possession of understandings or information that should be shared with the astronomers.

The information contained in this class of letter was typically of astronomical or cosmological concern. These individuals had gleaned the information they wished to communicate either by experimentation, observation or intuition and invariably felt a strong sense of urgency in their need to communicate their observations to the observers at Mount Wilson.

Letters of this kind began arriving at the observatory as early as 1911 and continue to arrive even today. There was, however, a swell in the letters received between the two World Wars, the years when the Observatory received, perhaps, its greatest public attention. During these years the letters were most often written to Milton Humason, Seth B. Nicholson, and Edison Pettit, all prominent astronomers of the time as well as Walter Adams who had assumed the role of Director of the Observatory from George Ellery Hale in 1921. In the 1940's the letters were collected and organized by Joe Hickox, chief solar observer. Since that time the letters have been passed from solar observer to solar observer, finally falling into the hands of Larry Webster, who most generously arranged for this exhibition.

LETTERS TO MOUNT WILSON OBSERVATORY

Auckland,
Tuesday July 7th

Dr Edison Pettit,
Dr Seth B. Nicholson,
Dear Gentlemen,
Some weeks ago I wrote
you a letter. Not yet having heard from you
I was wondering if you received my letter
I wrote you from Homai. Since, I have
shifted from Homai, to Auckland, so I
thought I would send you my new address.
I want to tell you I am not after money
& I am not a fraud. I believe I have
some knowledge which you gentlemen should
have. If I die my knowledge may die with
me, & no one may ever have the same
knowledge again. Because if people hear talking
they want stick, they go & do away with their
selves. I have gone through frightful things
still I go through it & I am beginning to
get knowledge. I would write down & tell
you what I no. But I would soon wait
till I hear from you. Because you are both
strangers to me & my letter may go astray.
When one writes one needs peace & quietness

CAT. # 0001

Letter from Mrs. Alice May Williams to Dr. Edison Pettit and Dr. Seth B. Nicholson

CAT. # 0001

Auckland,
Tuesday July 7th

Dr Edison Pettit
Dr Seth B. Nicholson
Dear Gentlemen,

Some weeks ago I wrote you a letter. Not yet having heard from you I was wondering if you received my letter I wrote you from Homai. Since, I have shifted from Homai, to Auckland. So I thought I would send you my new address. I want to tell you I am not after money & I am not a fraud. I believe I have some knowledge which you gentlemen should have. If I die my knowledge may die with me, & no one may ever have the same knowledge again. Because if people hear talking they want stick, they go & do away with their selves. I have gone through frightful things still I go through it & I am beginning to get knowledge. I would write down & tell you what I no. But I would sooner wait till I hear from you. Because you are both strangers to me & my letter may go astray. When one writes one needs peace & quietness

I have got half a house with another woman some years older. She will not let me sit quite a moment it is terriable she keeps wanting to no the inds & outs of everythink. She keeps running up & down the stairs in & out of the doors slamming them about & keeps wanting to talk & keeps wanting me to get ready to go out. It is awfull I dont no wether I am standing on my head or feet & still I am going through that treatment I told you. At times somethink works my mouth to talk out loud & I have got to be carefull of her hearing as she thinks I am mad, & makes all sorts of fun of me to people. So in a few weeks I may have a little house of my own, & in the meantime I may hear news of you people & then I will be able to sit down & write quietly without interference. you no yourself if people interfere with you you can't do your work properly. I do want to tell you somethink because the entrance into the other world is worked different to what you ever thought & you will get a shock. When I tell you I dont want no money from you. It wont do you no harm to have my knowledge. So I will now conclude hoping you gentlemen are living & in the very best of health as I hear that people are dying in america, with the very hot weather they are having.

I Remain
Yours Sincerely
Alice Williams
Mrs. Alice May Williams
No. 18 Norman Street
Rocky Nook off Dominion Rd
Auckland. NZ.

P.S. Please excuse writing & mistakes as this lady is worrying me to get ready to go out. Please keep my letters secret till I tell you what I no. Then you can do what you like. A.W.

Letter from Alice May Williams
to
The Observers at Mount Wilson

CAT. # 0003

Climate of the moon, absence of atmosphere. Great extremes of temperature, Can life exist on other Planets

Climate of the moon. The moon is a sphere and it works the clouds by night; it is not a Planet, & should not be interfered with.

The sun comes out by day and doses his work by day. He draws in the clouds and the sunspots you can see on the sun are that very black cloud, which is called a verandah, at certain seasons it is thrown out & spread over the sky for certain work going on underneath so it won't be seen, & to keep people from harm, on a very fine day it is drawn inside of one of those moonhouses to give a very fine day, so that is what a sun spot is. Black Clouds.

Some of those Planets you can see are like stations to do god's work. Venus is thickly blanketed with clouds, that is a place that stores the clouds. Jupiter be I suppose a storeplace for rain snow & hail. Others to keep air, wind, sunfire, wild fire, star fire, moisture, electricty, gases, It is worked by a human spirit world of human spirits that run over the top of the world & wind. The Planet mars is inhabited by human spirits like us can talk eat & drink wear clothes, but have great power. They are somethink people of this earth have never seen. They are kept to do work overhead. They also work our wireless gramphones, machinery, Moving Pictures Talking Pictures and all that sort of thing. All that sky is worked spirital. But all is gods workings of his worlds. Some time back I was asked if I knew what a rain bow was. I couldnt asnswer as I had not quite thought about it. But I think I can answer now. A rainbow is gathered up electricity & gases that could go off in lightening if allowed. I beleive the people on the Planet Mars have machenes that run down in the air elevater style. They can only come down a certain distance in the air & they look exactly like the moon. They have round glass shutters that look like the face of the moon. Inside they are rooms with beds tables chairs.

So I have been told & seen in half sleep trance. They are worked by three people We used to be on a lighthouse down south Island. I used to go outside a great deal at night. I no it was those machines come down in the air the funniest you have ever seen. hang right down close to earth in the air like a great big lamp I cant explain on paper. They run right up in the air & right down just like a elevater I havnt seen nothing like it in auckland When I used to go to bed The people used to show me inside their rooms. I think that is the reason they dont want people down on this earth to shoot rockets up into the moon. Because it might be one of those machines with people in and harm the people inside. I was told when I heard people talking & seen them in a trance to go up in a aeroplane as far as I could straight up in the air as one of those machines might be about & those people might try to get to me. One time I seen three brothers in one of those machines one looked like a clown they had like pipes set on a table in the centre of the room & they were alight like gas, a bed on each side. another time I seen a man sitting in a easy chair reading a paper & smoking a pipe he was dressed in a blue serge suit & had fair curly hair. Another time I seen two people a man & woman very tall fair people they where singing a moon song. They sung right through my mouth to show how they sing on gramphones. Another time I seen into a big room it had a glass case full up of lovely things I think I was told they were wedding presents. I have heard money jingle. All sorts of things just a few weeks ago. I had a sleep trance I was talking to some people I heard a child cry or talk then I took it by the hand & went down a winding road with like a big Terrace on one side. I seen table upon table of all lovely things to eat as if people had dined then left the table still laden with beatiful cakes & sweets lovely food I turned back took the child in my arm underneath. It started to kick up a row to be put down I put out my hand for a cake everything turned dark & I was like in a cloud. I woke up & I could still here talking going on. The next monday I had another sleep trance I came alive in a man arms he had his arms around me in a station kept for men coming off air machines he was dressed in wollen tights he took me outside and he left me to work on the other part of the station I no I was in another live world with a sky overhead it was night time with stars shining. like a star came down in the air. it looked like a air machine I seen them throw down somethink I run over to where it was falling it came out like a small parachute and it had somethink tied on it. It fell to the ground along side of me with a thud I picked it up it looked like a large box of sweets with scotch wrappers a girl caught hold of one and it all run out like milk or ice cream. The man told me to taste & I did so it tasted half like milk & ice cream I think it must of been their tin milk I seen them cutting up slices of tin corn beef. I picked up a small bag of new potatoes. The girl told me I made the father laugh as I was tasting all his stores & I tried to eat a raw potatoe. She told me that was the way they threw stores down to the men when they where going a long Journery to work in the air. That night I also seen two moons one half & the other some think I never seen before. I woke up I still heard talking. I have seen all sorts of moons & stars & openings. Tasted all sorts of foods & soups they let me smell all sorts of things show me having their meals. I have heard babies crying & screaming. I have had children talk to me & show me theirselves in a photo & show me pictures they think they are having a nice time talking to Al as they call me. Fancy a dear pretty little child of two years thinking he is great showing me his self to Al & having fun dear little baby young girls & men show me their selves like in a photo. like I am right awake show me their clothes they are wearing and making. how to make cakes sometimes I have only got to call through my gramphone & I get shown all sorts of things I Beleive the babies can hear me writing this. I think one wants to talk & show me hisself now. The children are nice. When I start reading someone reads it right out to me. When I go to the pictures Thoughs people acting in the pictures talk to me & yarn. Anyhow somethink talking inside of me enjoys it self at the Threatres & chats to me. So I am not lonely. But I do hate it saying who no, through me at times so people can hear, & it wont stop. It keeps working me up to say who no all the time, in the house, but not in the street my husband does carry on & he wont stop he thinks I am saying it on purpose. That is the reason they cant turn people on to hear & talk as somethink gets inside them & hurts them. I think you & Dr Pettit are the nearest in your estimations on Planets, are for heat & rain & snow. The rays of the sun is worked. Professor William Pickering computes that the new planet takes 656 years to make one circuit around the sun in an elliptical orbit. no such thing. let me tell you their are all different suns. It is to give nice warm days for pleasure & enjoyment to grow our plants & to keep us in good health. Do you mean to tell me you got the nice warm sun out in California overhead & the same sun out in the world of N.Z. & then Australia, they are all different worlds of this earth world. Do you mean to tell me, whole live worlds revolves around one sun in the air. It would be terriable. The live worlds of heaven are through the openings in the sky & you can't see them revolving I dont think so . The sky covers & the worlds run down & over. The waters that you see. May be the sea of the other world you can see through the holes of the sky. Those comets you can see are somethink that doses their work in the sky I should think so. Their is no other live worlds this side of the sky. But some of those large stars you see hanging in the sky may be sky machines with live people. But they can only lower a certain distance on account of those holes closing at different times & they may belong to different places & may have to wait a long time before they can get back to their own world.

I beleive their is machines called the magpie cages. They are exatly like a magpie cage with a large hole at the top. So I cant think of any more to tell you just now

I Remain Yours Sincerely

Alice M. Williams

My address
Mrs. Alice May Williams
246 Karangahape Rd. Newton
over Mr. J. Bells Furniture shop
Auckland

To Seth B. Nicholson
Carnegie Institution
Mount Wilson Observatory
Pasadena, California

Some of those machines are called sky planes. To build one of those machines you have got to build them to meet the air. The higher you get the greater the current of air. I beleive the people of the other world have glasses they can see you with. They can draw you to them.

The other live World
heaven

God made two live worlds, one he called heaven.
The people are made of heaven, Therefore heavenly
people. The other world he called earth, The people
are made of the earth Therefore earthly people.
One was made of heaven, & the other one of earth
Both similar to one another, I beleive heaven as all
different countrees like down here divided with
the sea, & a sky overhead. Between the two worlds
god put the sky to seperate the two worlds. The sky
is a covering to the other world. The sky is called
a firmament. That is the reason the people of heaven
cant talk to the people of earth. I beleive that sky
opens & closes on certain periods, When you see
all that cloud covering the sky right up & over.
Those clouds are called. Blinds, shutters & verandahs
Sometimes that sky opens underneath & you cant
see it on account of those clouds covering up the
sky. The sky opens slowly down like a lid of a
box then closes up quick. Then comes down again
& stops open for a certain length of time Then you
can see into the other live world right down & over.
I think it is this world the sky opens & closes I seen
it in a sleep trance. In a sleep trance I come
alive off my sleeping body onto my soul. To show
that when we die, we have another live body. Our
body down here is linked together, with a wireless
to our soul body in the other world. The spirit
of the body is our mind our thoughts. When we
die our mind our thoughts fly to our soul body.
& we come alive for our judgment The body dies
in this world & lives in the other world a
world god as set aside for us. He puts

CAT. # 0005

Letter from Alice May Williams
to
the Observers at Mount Wilson
CAT. #0005

The other live World
heaven

God made two live worlds, one he called heaven. The people are made of heaven, Therefore heavenly people. The other world he called earth, The people are made of the earth. Therefore earthly people One was made of heaven, & the other one of earth Both similar to one another, I beleive heaven as all different countries like down here divided with the sea, & a sky overhead. Between the two worlds god put the sky to seperate the two worlds. The sky is a covering to the other world. The sky is called a firmament. That is the reason the people of heaven cant talk to the people of earth I beleive that sky opens & closes on certain periods, When you see all that cloud covering the sky right up, & over. Those clouds are called. Blinds, shutters, & verandahs. Sometimes that sky opens underneath & you cant see it on account of those clouds covering up the sky. The sky opens slowly down like a lid of a box then closes up quick. Then comes down again & stops open for a certain length of time Then you can see into the other live world right down & over I think it is this world the sky opens & closes. I seen it in a sleep trance. In a sleep trance I come alive off my sleeping body on to my soul. To show that when we die, we have another live body. Our body down here is linked together, with a wireless to our soul body in the other world. The spirit of the body is our mind our thoughts. When we die our mind our thoughts fly to our soul body. & we come alive for our judgment. The body dies in this world & lives in the other world a world god as set aside for us. He puts us under Judgment. doses what he thinks fit for us But the world of heaven is a live world. I dont think any live Planet revolves around the sun. The sun is worked from day to day to do its work. You see sometimes a round place like the moon that works the sun fire all around. That place is one of the places to keep sun fire in to make the sun. It is worked spirital. I could show you myself if I was near you. havent you noticed in the day times at certain times you will see the sun on one side & you look in another you will see either a half moon or full moon in the day time. That isnt a planet. that is a spirit no planet can walk backwards & forwards over the sky. When the sun is drawn up high I could show you, that after the sun goes down it comes out like a great big star it is the sun drawn up high. You notice in the morning you will sometimes see a great big star like fire after a time you look in the same place & you will see the sun When the sun goes down that moon lights up to do its work to draw in the clouds or moister. You go out again in the early morning you will see the moon put out. Then if you watch it you will see it go right away & over & get out of sight. Now how can that be a planet or live world But I really think their are holes in the sky that open into the other world, & comes out like a moon some-times. Their are holes in certain places in the sky. You may be able to see certain countries of heaven through the holes. Thats what you think Planets. No live world this side of the sky. If I was near you I think I could teach you different things. I beleive their is a human spirit world. When it turns, you can get into the other live world, of human beings. It is very hard to write these things down as I have a lot of trouble & work to do at the present time. & to write these things down I have to work my mind & thoughts & memory. So I hope you can understand clearly what I write down here. Always from childhood I have been mad on Planets When I used to live on the lighthouse I used to live out side a great deal. & I watched everythink going on in the sky taught myself & also had a great deal of enlightment & that is mars, the human spirit world. That is what we call mars I think so. that is what I want to teach you gentlemen that everythink is different to what anyone of this world of today thinks. That sky is all different.

from Alice M. Williams

My address
Mrs. Alice May Williams
246 Karangahape Rd Newton
over J. Bells Furniture Shop
Auckland N.Z.

Dr Seth B. Nicholson
Mount Wilson Observatory
Pasadena, California

I have divided this into 3. I have put it in 3 different envelopes.

I will give you a design of one of their machines. They are round like the moon with airtight shutter at the front, glass. The machine must be built with some light material, airproof fireproof, waterproof. The inside must be like a room, 2 beds tables & chairs. Places to keep stores for 6 months or more. salt, must not be forgotten. flour, tin meat, potatoes, must not carry strong drink can carry money in gold or gold. must have condensers to condense the air outside, two long pipes like tubes inside one to let bad air out & one to let fresh air in must be worked elevater style to run up & down or along & accross. must have a small lavatory built in with a airtight sit & the door always kept close when not in use must not throw anythink into the air when the machine is in use, no slops of any kind. must have a tank built in for water for use while on the Journey Plenty fresh water, must not forget cooking utensils. something to cook with. At night time let up with a nice bright light. light attracts. If oil is used a large fireproof tank must be built to hold plenty of oil. Everythink must be built of the lightest of material. That all I can think of. A. Williams

The higher you get. The stronger the current of air & the more air running If those people from mars comunicated with people of this earth people wouldnt be able to do their work properly. They have got to put wireless on to your hearing I think that is what I got. It is terriable at times as somethink else interferes with their wireless & harms people. They can talk right down through my mouth like the gramophone no one else can hear them.

Letter from Andrew Plecker to Dr. Seth B. Nicholson

CAT. #0007

Sept. 7 1930.
Birmingham, Ma.,
2106 North 16th Avenue

Dr. Seth B. Nicholson
Mt. Wilson Observatory, Calif.

My dear Doctor:-

The very day of the discovery of Pluto, I wrote to Dr. Stetson, Perkins Observ. Ohio, with whom I was in active correspondence at that time regarding the real formation of our solar system, that the new body was not a planet but one of the incoming pieces of matter going to be a comet and then a planet of regular standing.

Our moon came in the same way 5,691 years ago and Mercury with the 1,000 Asteroids came about 900 years after our Moon. Venus and Neptune came first to our Sun; then came our Earth 113,931 years ago. This we get from the study of history and chronology as kept by the nations before our Moon caused the destruction usually called the universal flood, 3,761 B.C.

Pluto is as yet no planet and therefore has no moon or moons. There may be more than one piece, yea 1,000 pieces, (see above) and they may travel together but there cannot be a relation between them which we call lunar.

Before the "Coming of our Moon" our Earth had no 24 hour day, but a day, "The Long Day", which lasted 25,920 years. This we see in Chinese, Hindu, Aztec and Maya chronology. At that time people wandered with the Sun, their God of the Garden, and there were nights and winters 12,920 years long, yet the people had "Everlasting Light" on account of wandering.

Kindly let me hear from you.

Very sincerely, Andrew Plecker

I wrote a book about this 15 years ago but no one would publish it. A planet without a moon does not axiate rapidly.
I gave this to Lick Observ. some 12 years ago.

Letter from an Unknown Person to the Observers at Mount Wilson

cat. # 0008

The creation of God or Astronomy

First I wish to say for the benefit of some who ridicule the Idea of there being other worlds besides this one on which we live, I once was takeing a minister of the gosple to his apointment to preach, it was on sunday morning and as I spake to him of the wonderfull creation of God how God had Created millions of other suns and worlds, says he what do you mean by saying other suns and worlds, I answered by saying I mean the other suns and systems of worlds, for God no doubt would not make suns to shine without each sun had a system of worlds such a sun would be to no purpose it would not be of any use to any one not even to its Creator one had Just as well say I am going to build a big costly mansion but I am not going to let any one live in it, the mansion would be of no use to any one Just time and material and energy Just a wast of time and no good done, so no doubt from a reasoneble standpoint, each sun has its system of worlds and the worlds of each system are for the purpose of habitation for Gods inteligent creatures to lie on and be hapy. Through all time to come: then he sais to me is there an other world and where in the Bible do you find where it speaks of other worlds: well says I to him the Bible says in many places that God made the worlds and Jesus said himself that he mad the worlds St. John 1-3 all things were made by him and many other bible texts prove that God has made the worlds by millions and yet the minister did not believe that there were any other world only this one on which we live and it as flat as a pancake round as a pancake but not round as a ball and that the sun did actualy go around this pancake every 24 hours Just because God did not explain every thing in his word: the minister thought that because Joishua comanded the Sun to stand still, and so it did yes and every planett in this solar system also stood still if it had not this earth would of bin an hour behind and thousands of miles from its Orb Joshua was a god man his heart was right but his head was wrong God Just answered him acording to what he wanted and not what he said: it is very grievious to me to know so many bible students of today that might themselvs on the creation of God and they dont believe that there is other worlds and other solar systems Just because it is not explained in the word of God: if I only had the means to procure a scope of 3 or 4 inch lens I would go back to Illinois and convince some of them unbelievers and not only acumulate quite a sum of money at 10 cts a sight but convince a number of gainsayers and if I could learn of any one that has a good carrying scope that could be used on the streetes to exibit and show the stars Moon and sun I could earn not only for my self but a quite a sum for the man who lent me the outfit or if arangements could be made to make payments on susch an outfit I could pay for the outfit in due season. I am well satisfied, I have bin reading up quite a good deal of late since I have came to Calafornia on astronomy and hold it as a study next to my Bible for it is God's creation and worthy to be studied as all God loveing people should. For I never have seen an Infidel astromer I think they all are true believers those that have viewed through the scope. I never have seen through a scope but once on top of the 22 story Masonock Temple in Chicago a view of the sun with a 4 inch scope and then only a minit and one other time saw a few stars through Dr Youngs scope here in L.A. I would believe if I never had looked through a scope for I believe that if we could travil a hundred of Decillions of light years from here that there would still be systems of rushing worlds for God never had a beging and never was Idle and is every where present and is the only Creator there is no other being in the entire space of all imencity that can createt but God alone no other being can bring something out of nothing but the Father Son and Holy Ghost and had plenty of time to create all that is in the Stelar regons of imencity even if he had only created one grain of sand each Decillion of years: and then have plenty of time to travil a Decillion of milkey ways and yet not be halfway to another stelar reagon, for as we know they might be a Decillion of milkey ways so far from each other that a streak of lightening would have to travil a thousand years from one to the other I have writen a book entitled (Will we know each other in heaven) I also have some astronomy in the book some of my friends thought I had better of left the astronomy out as that was God's buisness and that I or no one elce knew what the stars is some of them think that they are only little sulpher balls up in the sky and that is all they are but if I could only have a scope to convince them of theyr ignorance I should be Glad. as well as to benefit and acumulate finance if God did once not exist, then nothing did exist then nothing could not of taken nothing and mad something so by that he proves himself to be a self existing God without begining or comencement.

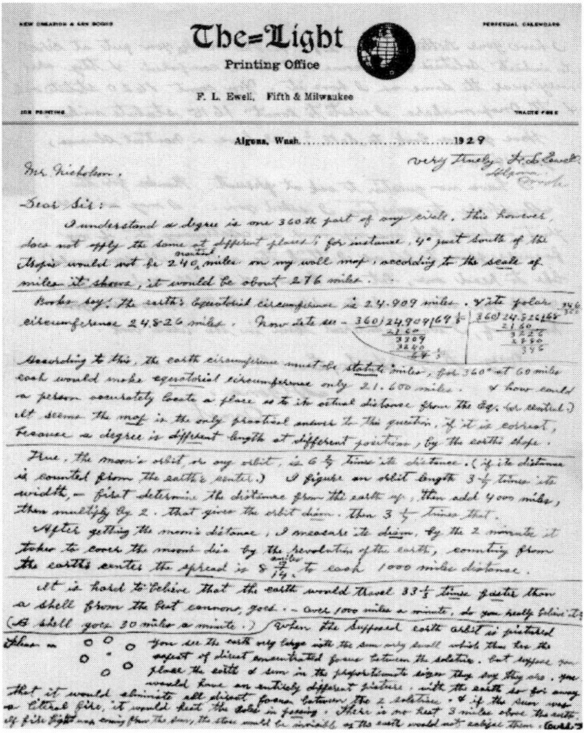

NEW CREATION & NEW BOOKS — **The=Light** — PERPETUAL CALENDARS

Printing Office

F. L. Ewell, Fifth & Milwaukee

JOB PRINTING — TRACTS FREE

Algona, Wash. 1929

Very Truly, F. L. Ewell, Algona, Wash.

Mr. Nicholson,

Dear Sir:—

I understand a degree is one 360th part of any circle. This however, does not apply the same at different places; for instance, 4° just South of the Tropic would not be 240 nautical miles on my wall map; according to the scale of miles it shows, it would be about 276 miles.

Books say, the earth's Equatorial circumference is 24.909 miles. & its polar circumference 24.826 miles. Now lets see — 360)24.909(69 1/5 | 360)24.826(68

According to this, the earth circumference must be statute miles, for 360° at 60 miles each would make equatorial circumference only 21.600 miles. & how could a person accurately locate a place as to its actual distance from the Eq. (or [illegible]) It seems the map is the only practical answer to this question, if it is correct, because a degree is different length at different positions, by the earth's shape.

True, the moon's orbit, or any orbit, is 6 2/7 times its distance. (if its distance is counted from the earth's center.) I figure an orbit length 3 1/7 times its width, — first determine the distance from the earth up, then add 4000 miles, then multiply by 2. That gives the orbit diam. Then 3 1/7 times that.

After getting the moon's distance, I measure its diam. by the 2 minutes it takes to cover the moon's disc by the revolution of the earth, counting from the earth's center the spread is 8 9/14 miles to each 1000 miles distance.

It is hard to believe that the earth would travel 33 1/3 times faster than a shell from the best cannons goes. — over 1000 miles a minute, do you really believe it? (A shell goes 30 miles a minute.) When the supposed earth orbit is pictured thus — [diagram] you see the earth very large with the sun only small which thus has the aspect of direct concentrated focus between the solstices. But suppose you place the earth & sun in the proportionate sizes they say they are, you would have an entirely different picture, with the earth so far away that it would eliminate all direct focus between the 2 solstices. & if the sun was a literal fire, it would heat the [illegible] in passing. There is no heat 3 miles above the earth. If fire light was coming from the sun, the stars would be invisible as the earth would not eclipse them. (over)

CAT. # 0009

LETTER FROM F. L. Ewell
TO
THE OBSERVERS AT MOUNT WILSON

CAT. # 0009

1929
very truly F. L. Ewell
Algona Wash

Mr. Nicholson,

Dear Sir:-

I understand a degree is one 360th part of any circle. This however, does not apply the same at different places; for instance, 40 just south of the tropics would not be 240 nautical miles on my wall map, according to the scale of miles it shows, it would be about 276 miles.

Books say, the earth's equatorial circumference is 24.909 miles. & its polar circumference 24.826 miles. Now lets see-

According to this, the earths circumference must be (statute) miles, for 360 at 60 miles each would make equatorial circumference only 21.600 miles. & how could a person accurately locate a place as to its actual distance from the Eq. (or central.) It seems the map is the only practical answer to this question, if it is correct, because a degree is different length at different positions, by the earths shape.

360 / 24.909 \ 69 1/5
2160
2160
330
69 1/5

360 / 24.826 \ 68
2160
3226
2880
346

346/360

True, the moon's orbit, or any orbit, is 6 2⁄7 times its distance, (if its distance is counted from the earths center). I figure an orbit length 3 1⁄7 times its width,- first determine the distance from the earth up, then add 4000 miles, then multiply by 2. That gives the orbit diam. Then 3 1⁄7 times that. After getting the moon's distance, I measure its diam. by the 2 minutes it takes to cover the moon's disc by the revolution of the earth, counting from the earth's center the spread is 8 9⁄14 miles to each thousand miles distance. It is hard to believe that the earth would travel 33 1⁄3 times faster than a shell from the best cannons, goes.- cover 1000 miles a minute, do you really believe it? (A shell goes 30 miles a minute.)

Plus

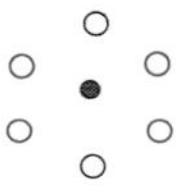

When the supposed earth orbit is pictured you see the earth very large with the sun very small which thus has the safest of direct concentrated focus between the Solstices. but suppose you place the earth & sun in the proportionate sizes they say they are. you would have an entirely different picture. with the earth so far away that it would eliminate all direct focus between the 2 solstices. & if the sun was a literal fire, it would heat the poles in passing. There is no heat 3 miles above the earth. if fire light was coming from the sun, the (sto -) would be invisible as the earth would not eclipse them. (over). I have your letter of Nov. 27. & the marks you put at cir(cls) to indicate solstices & Tacoma. I have compared and they are very near the same as I have it. You count 1620 statute miles & the map makers I wrote to count 1615 statute miles. Have you a book to sell? I have a Nautical Almanac, year 1930. I have no question to ask at present. Thanks for the Answers to questions I asked you. I may write and print a small book on my views on Astronomy, etc. if my focuses prove to be as I think they will. If you would like to read one, let me know, & I will send one free, & if you will send the address to all the Observatories you know of, I would interest them in the little book.

Very Truly
F.L. Ewell
Algona,
Wash.

LETTER FROM W. L. BASS
TO
DR, SETH B. NICHOLSON

CAT. #0011

August 20, 1929

Dr. Seth B. Nicholson
Mount Wilson Observatory
Pasadena, California

Dear Sir:-

As the work of propagating my Theory of Celestial Growth has progressed to the stage of history in the making, I purpose including in its composition, records of my contemporaries who are competent to be incorporated in this chronicle.

To this end I request that you will be good enough to favor me with your photograph - autographed, if you will be so obliging - as my plan is to present to all who are concerned, the view, likenesses and signatures of scientists representing authority on the subject of Cosmogony, as recorded in my era, especially the year 1929.

If you have no photograph at hand, and will advise me as to who your photographer is, authorizing me to apply to him for one, I will purchase it and appreciate your permission to do so.

Very truly yours,
W. L. Bass

WLB:GVW W.

Letter from W. L. Bass
to
Dr. Seth B. Nicholson

CAT. # 0012

October 11, 1929

Professor Seth B. Nicholson
Mount Wilson Observatory
Pasadena, California

Dear Sir:-

I am developing a work on Cosmogony and desire to include in the records individual views of astronomical scientists of the present period. My plan is to publish these opinions, and realizing that anyone entertaining profound convictions will be willing to sponsor whatever reply he accords to ensuing matter, I request that you will favor me to this extent:

(1) Do Mercury and Venus - alleged to lack satellites - execute the <u>precession of the equinoxes?</u>

(2) Does Luna, in its relation to Earth, execute the identical function as that which Earth performa in relation to Sol: <u>precession of the equinoxes?</u>

(3) Do the numerous satellites of Jupiter and Saturn, in relation to their respective primaries, execute the identical function as that which Earth performs in relation to Sol: <u>precession of the equinoxes?</u>

Anticipating a response, I thank you.

W.L.Bass

WLB:GVW

2010 N. Park av.
Philadelphia,
Pa.

Dear. Dr. Adams:—

Unfortunately for Astron-
omy. the great expenditure
and the ridicule for its pay;
while the mighty sun looks
down upon many serious
mortals anxiously search-
ing for the cause of a
strange visitor. Eclipse.
Dear. Gentlemen;

CAT. # 0013

Letter from Mrs. Helen Hartman
to
Dr. Adams
CAT. # 0013

2010 N. Parkam
Philadelphia,
Pa.

Dear Dr. Adams:-

Unfortunately for Astronomy, the great expenditure and the ridicule for its pay; while the mighty sun looks down upon many serious mortals anxiously searching for the cause of a strange visitor. Eclipse.

Dear Gentlemen, there are no shadows in connection with the eclipse.

It is ________________

At your service
in the name of justice.

Mrs. Helen Hartman

CAT. # 0014 *front*

POST CARD
NOTIFY YOUR CORRESPONDENTS OF CHANGE OF ADDRESS
MADE IN U.S.A.
FOR CORRESPONDENCE
FOR ADDRESS ONLY

MR WILSON
PLEASE SHOW MR EINSTEIN
YOUR BIG TELESCOPE SO HE CAN
TELL US ALL ABOUT IT HE HAS
NO BIG TELESCOPE YOU KNOW
BUT WE KNOW HE IS A BIG SIENT
IFIC MAN IN EDUCATION HE IS
CONSIDERED EVEN GREATER
THAN CHREY CHAPLIN
HURAY FOR THE JEWS WE WILL
SOON RULE THE WORLD SOME
FELOWS DO NOT LIKE US BUT
WE GOT THE MONAY HURAY
HURAY

SERIES 1206 A

Mr Wilson
Mount Wilson
OBSERVATORY
SOUTH CALIFORNIA

CAT. # 0014 *back*

Postcard From Unknown Person
to
Mr. Wilson of Mount Wilson Observatory

CAT. # 0014

MR. WILSON
MOUNT WILSON
OBSERVATORY
SOUTH CALIFORNIA

MR. WILSON

PLEASE SHOW MR EINSTEIN YOUR BIG TELESCOPE SO HE CAN TELL US ALL ABOUT IT HE HAS NO BIG TELESCOPE YOU KNOW BUT WE KNOW HE IS A BIG SCIENTIFIC MAN IN EDUCATION HE IS CONSIDERED EVEN GREATER THAN CHARLEY CHAPLIN HURAY FOR ALL THE JEWS WE WILL SOON RULE THE WORLD SOME FELLOWS DO NOT LIKE US BUT WE GOT THE MONAY HURAY

HURAY

READ—THEN PASS THIS TO A PROPER PERSON

THE EARTH is FLAT and STANDS FAST. PROVE IT

Delusions.

Delusions, or fictions in some cases have become public opinions—as in the case of the shape of the earth.

I dreamed last night I was in a court—when a man arose and made a charge against me. No action against me being taken by the judge—I followed my accuser from the court room and outside—I asked him what is the trouble?

He said—you told my hired man that the earth was flat and stands fast. He being an educated foreigner made a drawing to show your plan of a flat earth. Then he took a copy of the plan and began to fasten it to the out-side wall of a building. I looked at the drawing and I said it was a good work—let it stay.

My accuser then took the copy of the drawing from the wall and threw it on the earth. I picked it up and took it into the court and showed it to the judge—and he saw there were two copies. I then took my seat in the court in peace.

I also had a dream that I had been making new earth from waste materials including the flesh and grease of animals. The pile heated and burned—and the surface gave way in places and smoke came up.

Thus the earth—as a whole—in places was made.

Aug. 29, 1920. Historian, Boston, Mass.

Over

CAT. # 0015

Letter from Unknown Person
to
the Observers at Mount Wilson
CAT. # 0015

READ - THEN PASS THIS
TO A PROPER PERSON

THE EARTH is FLAT and
STANDS FAST. PROVE IT

Delusions

Delusions, or fictions in some cases have become public opinions - as in the case of the shape of the earth. I dreamed last night I was in a court - when a man arose and made a charge against me. No action against me being taken by the judge - I followed my accuser from the court room and outside - I asked him what is the trouble?

He said - you told my hired man that the earth was flat and stands fast. He being an educated foreigner made a drawing to show your plan of a flat earth. Then he took a copy of the plan and began to fasten it to the outside wall of a building looked at the drawing and I said it was a good work Let it stay. My accuser then took the copy of the drawing from the wall and threw it on the earth. I picked it up and took it into the court and showed it to the judge - and he saw there were two copies. I then took my seat in the court in peace.

I also had a dream that I had been making new earth from waste materials including the flesh and grease of animals. The pile heated and burned - and the surface gave way in places and smoke came up.

Thus the earth - as a whole - in places was made.

Aug. 29, 1920

Historian Boston, Mass.

The builders of Greece came from a country west from Egypt beyond the ocean - which of course was America. The authors of the Bible tell us that the father of Abraham came to Egypt or Palestine from over the flood - which means that the father of Abraham came from a country west from Egypt beyond the ocean - which of course means America. The father of Abraham mated with a woman or women of Egypt or Palestine from whom came the Israelites in part at least. The Hindus in the Rigveda tell us about Heaven - that was beyond the dawn. This means that to the Hindus in India - Heaven was beyond 105 or 120 degrees of longitude east from India - as the dawn means east from any place - and in the Rigveda the place was India. 105 or 120 degrees of longitude east from India would be a long way to Heaven - now called America.

Letter from George P. Tobias to Dr. Milton Humason

CAT. # 0016

Aurora Ill Jan 21 1932
Dr Milton Humason
Mount Wilson Observatory Calif

Dear Sir:

Please pardon my audacity in interrupting you in your work. I Will be as brief as possible in trying to make myself clear.

I am writing a thesis based on an entirely new angle of the creation of the Universe.

This thesis is of a philosophical or scientifical nature rather than astronomical, of which the following extracts will give you an idea of the theory, all of which can be verified by demonstration or by existing facts:

"It is impossible for any human mind to comprehend the vastness of space. Stretching on and on without limit. No beginning no end in all directions from any point, there is always more space beyond.

We are also conscious of the existance of power, an all mighty power which is everywhere and is responsible for the creation of everything. The only power that is capable of creating something from nothing. The whole Universe and everything pertaining thereto was created out of absolutely nothing. This power is in all things. The all in all, and without it nothing could exist which does exist. It built the Universe, the most gigantic, the only perfect piece of machinery in existance. Forever in motion.

All space is thoroughly saturated with this mighty power, which is magnetism sustained in etheral atmsosphere. This Universal magnetism appears in various degrees of voltage units in all things according to the mass of the body. The more mass the greater the voltage, possessing various stages of attractive power, forming an etheral atmostphere in addition to an electrical or magnetic atmosphere, which envelope and protect the nucleus and no matter from the outside can come within this magnetic atmosphere until it becomes depolarized. In other words all outside matter is held in quarantine until the nucleus absorbes the polarization of such matter.

A disturbance in the etheral atmosphere causes solid matter to form in fine dust becoming larger, gathering into clouds, relatively magnetized particles assemble forming a nucleus around which gather still more particles finally resulting in a large cloud of nebula. There are always two varieties of dust in these clouds, light star dust which is luminous and create stars or suns, the dark non-luminous goes into planets. Only the light particles are visible to us. Sometimes, when nucleus is shown edgewise to us we notice black streaks, this is the dark dust eclipsing the light particles. As the star is forming, simultaneously planetary orbits are being formed. The electric pencils shooting out in all directions and from all points on the star cross at points tangent or focal points, in spherical form, a number of times throughout the entire electrical atmosphere. The nodes becoming wider and wider between the focal points or orbits as the distance increases from the star, these focal points are the only points where heat is concentrated. In the orbits planet dust is assembled, large and fine particles. This will eventually be gathered and form a planet.

All bodies suspended in space weighs nothing. The power exerted by the sun in causing them to revolve and cary them forward in their orbits is almost negligible. A person, given a footing, could perform the task with one finger. Of nature, all stars and planets are as cold as space itself and will remain so until heat is transmitted to them by electricity.

The sun is not a ball of fire, quite the contrary, it is no hotter on the face of the sun then it is on earth. Nor is it hotter on Mercury than it is on Neptune.

The earth has a definite motion from north to south. Pole over pole, in a spiral course. There is probably no spot on earth that has not been covered by the arctic or antarctic circle, and there is no part of the earth which has not been traversed several times by the equator.

Comets are the chore boys of the Universe. Constantly on the job. Feeding planets and stars and they are composed of star and planet dust, large and small, all solid matter. Radium and light material going to the sun and the dark to the planets. Our mountains were given to us by comets encircling the earth in the manner which is at present in progress on Saturn, Jupiter and Uranus. Over a period of thousands of years the mountain ranges all over the globe, including those in the seas, were deposited. Prior to this there was plant and animal life, including human beings on earth; but this operation wiped out all forms of life. There was an almost continuous rain of rock, ice, clay, salt and metals over all parts. Fish and amphibious animals were accustomed to fresh water, the large quantities of salt thrown into the waters killed them. Dust to the thickness of several hundred feet covered the portions between the mountains extinguished all life.

When the Arctic circle covered points in the Pacific, southwest from central America, the equator crossed the American continent at what is now north and south. Between Greenland and Alaska we find traces of temperate and torid zone animal and plant life. A thousand years ago Polaris was exactly north, it is now west form north, this indicates that we are traveling slowly nearer the equator.

This is but a jumbled mess hitting the high spots; But you can gather enough to give you an idea of the theory.

If you are bored by the foregoing bunk just dump it into the waste basket, but if interested would be pleased to hear from you and will make an effort to explain further. Would also be pleased to have the opinion of Drs Einstein and Dr. Sitter.

Thank you,
Geo P Tobias
384 Spruce St
Aurora Ill

Detroit Mich
Nov 3th 1933
A Cycle in the sky
Dr. Seth. B. Nicholson

Your honor which I
Know is true between
11 & 12 A.M. I just happen
to look up and I
sighted the cycle with
my nacked eyes when
seen at frist it was
at tach to the white
cloud that surounded
the sun the color of it
was as a lamp light
in day time the handle
of the cycle was turn
south west the blade
drown to the earth the
point north east I
spoke to my family
about it and a few
friend they said it
either ment

CAT. # 0017

Letter from Mrs. H. Hesson to Dr. Seth B. Nicholson

CAT. # 0017

Nov 3th 1933
Detroit Mich
A Cycle in the Sky
Dr. Seth. B. Nicholson,

Your honor which I know is true between 11 & 12 A.M I just happen to look up and I sighted the cycle with my nacked eyes when seen at frist it was at tack to the white cloud that surounded the sun the color of it was as a lamp light in day time the handle of the cycle was turn south west the blade drown to the earth the point north last I spoke to my family about it and a few friend they said it either ment reach unto the north west an to the north east and the dark.

Sincerely

death or trouble that was to weeks ago and about a month befor the cycle there were to paths in the skys during the midnight hours nothing could not be seen in the skys but the to paths the white path with a strip of dark clouds on each side look to be wide enough for 3 persons to walk side by side in it which layed away in the north an the length there of reach unto the north west an to the north east and the dark.

Sincerely

path way with white strips of clouds the length of it was like wise and wide enough for four cars to go down it side by side it look to be many miles this side of the white path I said to my children which path must we choose they said the white path now I leave it to you to asked the world which path will they take to show good will twas all man kind

Sincerely Yours,
Mrs. H. Hesson

LETTER FROM BOBBIE MERLINO
TO
THE OBSERVERS AT MOUNT WILSON

CAT. # 0018

129 Georgia Ave.
Atlantic City, J.J.
Dec. 4, 1932

Dear Sir:

I am a lad of 18 years of age. I appeal to you for one reason. Since childhood I've always been interested in planets especially the Red Planet Mars. So for this reason I write to you for inquiries on, will there ever be an interplanetary expedion to Mars in the near future. I would like to accompany that expedition. For this reason I ask you because through your medium and reference I could easily be one of the accompaning aides for the cause of science.

I want to reveal that innermost secrets of Mars which are puzzling the scientists the world over. I will beleive and always beleive that the Planet Mars is inhabited. But not in the same stature and conditions of you, me or anyone else. I've always wanted to visit that planet in the early future. I readily understand that is a very dangerous expedition that we may never return but as long as I just take one glimpse at it I am satisfy if die on the Planet I've always planned to visit. I am not out of my mind. I am as sane as anyone and I am very serious about this matter. Hoping you take this serious, I remain

Yours truly
Bobbie Merlino

P.S. In case you want to respond here is my address plainly.

Bobbie Merlino
129 N Georgia Ave.
Atlantic City, N.J.

LETTER FROM P.O. Argle
TO
Dr. Fredrick E. Wright

CAT. # 0019

Dec. 1st 1930
Dr. Fred E. Wright
astronomer
Washington D.C.

Dear Doc Wright-

I notice an article in Pathfinder of Nov. 29 page 28 stating that astronomers of the American Moon (Committee) having finished their studies and the new theory of pumice stone is (—) on reflection tests- yes-but this is not saying that the Moon is really pumice stone but merely affirming the theory- but a theory is not always a fact- however I have been affirming that the Moon is (mearly) a kind of pumice stone, and have taught this theory for about 20 years. So the theory is not new to me.

I am a freelance scientist. I teach that the Moon is a part of a core from a busted up earth or planet as you may say. It is most likely a part of this earths core. as our earth was once busted or blown to pieces. We have plenty of evidence of this. Without turning to Holy Writ. but Holy Writ affirms this theory without a doubt.

The center of the earth is a solid living coal of fire. since if a part of it should be busted off and blown out it most likely would (—) off and would appear as a kind of pumice stone.

The earth was once much larger than it is at present and it became (—) a large amount of it flew off into space the balance of it merely crashed together and resumed its journey. Of course this catastrophy changed the shape of the earth and probably changed its course and possitions somewhat. The chunk of core being near the center it did not get force enough to blow it very far and having no mineral weight it could not settle back on Earth. Hence it is mearly held at a (—) distance from the earth (—) by gravity of light.

The Moon is a great reflector of light, but a sorry condenser. It being a soft substance it does not catch or hold light the heavy light mearly passes through it while the soft and bright light glances off. This is why we get no heat from the moon that amounts to anything. This soft reflected light is not the heating sort light is in two parts male and female. the male produces no heat, but when the male and female is condensed together they become the hottest of all the heats. to give you a good idea of how they work I will have to explain the working of the sun.

The sun does not make the light. but it is the light that makes the sun. light passes to and into the sun. and is condensing all the time. the more it condenses the brighter it gets. It meets other rays coming in from every direction. These rays meet at the center and explode. (these) waves are driven out one after another all the time these waves are (glob—) bubbles one pressing against another they (keep) going and enlarging all the time. they go for billions of miles until they strike other waves of light coming from other suns. These waves crash head on like two bulls a fighting. they are busted up and the pieces which (then) then spin out into slim streaks passing back to the sun (—) and condensing again over and over all the time (—) for the fraction of a second. this is why the sun never can burn out. light won't burn. Light enters and passes through the hottest of heat and coldest of cold without being effected Light consumes nothing and nothing consumes it yet nothing can live without it light is in the life of everything it changes things but does not consume them. Light is eternal. when those waves of light strikes the Earth they are broke up into atoms then atoms whirl at great speed and the friction (cause) the heat which is supposed to come from the sun but really it is only friction from the light (coming) in. All of it doesn't (—) in but a great (—) does. There is nothing that light cannot penetrate or pass through. Some things very fast, some slow and some very slow. After light enters the atmosphere of the Earth it can be captured. various minerals and chemical compounds catch it and hold it sometimes for a long time. Then again the (—) become (over-) and blow up- when light strikes the Earth the waves is broken up and the atoms whirl and fly in every direction some (—) in while some glances off and passes out through space. others go direct to the sun light that enters the earth passes very slowly until it reaches the core when it condenses and is driven out by its own pressure. and while (—) out it after meets light coming in, and if these particles meet dead center they (glue) together and the coming in light passes back and outwards and is sometimes captured or detained until force is gained from (more) light coming (—) it breaks out and we see it as lightning Man can make lightning by catching light with his machine. and chemical batteries. But he calls it Electricity. Light condensed in the Earth and while passing out of it be cought and held by (—) minerals it becomes molecules. Molecles produce life of all sorts according to their composition and strength. They also furnish the explosive power in Gas, Steam, Gunpowder of every class. They differ from Electricity in they (posess) in a globular form in every direction. While Electricity shoots off in a narrow streak under high pressure in a spiral (—), under law and natural (—) it is straight Condensed Earth light can be (—) of as Aurora Borralis. Light striking the Earth produces a great amount of heat. this heat is what pushes the Earth around. simply heat and cold is what does the job. The light whirling up from the Earth in every direction is what causes Gravity. All things have weight according to their density, the heaviest sinks lowest towords the center this is why the moon cannot come back to Earth. It is of a very light material hence it merely floats at a (—) distance. Gravity works both ways upwards and downwards a block of stone will sink in water while a block of wood the same size will rise. Same way in the air and light. but you understand this I mearly mention it. to show that the moon is of a very light substance. Pumice stone will swim in water (—) if a large chunk be suspended above the Earth and (—) with light it follows that it will float just like a gas bag floating in air. (M) Dr. Wright, I have wrote you somewhat a lengthy letter. But I could not explain the workings of light in connection of the Moon with fewer words. less you might not understand. but I could have used many more words, but this might annoy you- as I don't know wether you are interested with my ideas or not. But I do know they are new or at least they have never been published. However if you are not offended and wish to know more. why just write me and I will tell you all you wish to know about light I have no object in addressing you except to let you know that I know something about light that others don't seem to know. If they do they want tell it. But if a man knows something that others ought to know then he ought to be willing to tell what he knows Its of no use to write a letter like this and send it to news papers or magazines (—) they wont publish it, or pay any attention to it at all. all persons writing for the press must be (—) or backed by some well known society or he gets no hearing

Yours respectfully
P.O. Argle
of the Am. Academy of Science

Liberty Tenn.

LETTER FROM F. L. EWELL
TO
SETH B. NICHOLSON

CAT. # 0020

Feb 6, 1932

Seth B. Nicholson
Pasadena, Calif.

Dear Sir;-

Do you know of an Observatory on or near the Equator or Tropic of Capricorn,? I would like to get such address. There is a dispute as to what shape the moon orbit would discribe going around the sun, & also around the earth the same time. Would you make a rough sketch of what shape it would be.

There is another argument here, about the sun rise and sun set. One man says the sunrise at exactly 6.a.m. & 6.p.m. set, on the Equator, at Equinox time of year, is the top edge first seen at rise & last seen at set. The other party says it is the center of the sun seen at that time. Who is correct about this? Did you ever see it? Who did?

Very truely,

F.L. Ewell
Rt, 3, Box 330-P.
Auburn, Wash.

WESTERN UNION
TELEGRAM

0D6 K DVA192 NL PD=DENVER COLO 15=

THE WORLDS LARGEST TELESCOPE=
PALOMAR MOUNTAIN CALIF=

GENTLEMEN ARE YOU INTERESTED IN SEPERATING VALUABLE CHEMICAL COMPOUNDS FROM THE SUNSHINE RAY? WORTH BILLIONS OF DOLLARS APPRECIATE AN AIRMAIL REPLY=

CHARLES V CARROLL REPRESENTATIVE SUPREME CREATION TREASURE CARE US NATL BANK DEPT FOREIGN TRADE DENVER COLORADO=

(10) AMP

CAT. # 0021

Telegram from Charles V. Carrol
to
the Observers at Mount Wilson

CAT. # 0021

OD6 K DVA 192 NL PD= DENVER COLO 15= THE WORLDS LARGEST TELESCOPE= PALOMAR MOUNTAIN CALIF=

GENTLEMEN ARE YOU INTERESTED IN SEPARATING VALUABLE CHEMICAL COMPOUNDS FROM THE SUNSHINE RAY? WORTH BILLIONS OF DOLLARS APPRECIATE AN AIRMAIL REPLY=

CHARLES V. CARROL REPRESENTATIVE SUPREME CREATION TREASURE CARE US NATL BANK DEPT FOREIGN TRADE

DENVER COLORADO=

(10)AMP

Long Beach Cal
1237 East 2nd Street
July 27 1948

Observatory
Dear Sirs
Gravitation is not
a mystery as taught by lying
scientists but contemptible
lying
Not a scientist nor an
Educator, nor an astronomer
seeks the truth about
gravitation but always
seeking some lying excuse for
backing senseless gravitation

CAT. # 0022

Booklet Prepared by T. P. Stanley and Sent to the Observers at Mount Wilson

CAT. #0022

Long Beach Cal
1239 East 2nd Street
July 27 1948

Observatory
Dear Sirs

Gravitation is not a Mystery as taught by lying scientists but contemptible lying

Not a scientist nor an Educator nor an Astronomer seeks the truth about gravitation but always seeking some lying excuse for teaching senseless gravitation as the great force of the Universe

Only lying Astronomers will teach the Law of Gravitation and explain that particles of matter are too small to show their attraction for each other.

Only cheating Astronomers will teach that "All bodies attract each other by gravitation" when not an astronomer can find one body attracting another body by gravitation.

When astronomers were discovered lying in teaching that large heavy bodies were falling faster than small light bodies of same density according to the law of gravitation the lying scientists made a new lying law of gravitation that "All bodies fall in the same time under gravity alone" which is in complete contradiction of Newtons lying law of gravitation. Yet Astronomers are willing teachers of both lying laws of gravitation in which truth and sense can not be found.

When astronomers were discovered lying in teaching that falling bodies were caused by gravitation in the falling bodies and in the earth the lying scientists taught that all the attractions of gravitation were located in the center of the earth to show why no attraction of gravitation can be found although taught in lying textbooks to be everywhere in all bodies.

Lying Astronomers teach that gravitation makes bodies fall in a straight line from their centers of gravity toward the earth's center of gravity which is very stupid lying for all bodies are falling in a curve toward the center of the earth because all falling bodies are revolving with the earth which is not turning under any falling body.

Lying astronomers teach that gravitation makes a 150 pound man on the earth weigh only 25 pounds on the moon while on Jupiter the man would be crushed with his enormous weight showing the lying foolishness of astronomers in teaching that gravitation produces weight.

Astronomers were discovered lying in teaching the moon attracted the water of the oceans into high tides by gravitation when it was learned that high tides are on opposite sides of the earth at the same time so the lying scientists taught the moon attracted the earth from the water to explain the opposite high tide although the earth is fifty times larger than the moon showing how willing astronomers are to teach plain falsehoods.

What is it about science which makes willing lying astronomers who were discovered lying in teaching the sun attracts the planets and the moons by gravitation hence another lie by scientists was required to teach an opposing centrifugal force which can be found with air resistance on the earth, yet another lie by scientists was required to explain the Eclipse in which the planets were revolving around the sun.

Astronomers are willing to lie in teaching the sun sends heat down through sub zero air to the earth which is absolutely impossible. No scientist ever wrote a truthful article about gravitation which all scientists call a mystery because they refuse to see the lack of common sense and truth in teaching gravitation which has no existence.

Falling bodies and weight under the force of the downward air pressure as shown by the Barometer is too much like common sense to appeal to scientists who prefer to teach that falling bodies and weight are caused by senseless gravitation for the lying scientists know that no one can find the mysterious disappearing attraction of gravitation.

Electricity is shown in waves producing high tides in short waves with wave lengths of air pressure to cause the surf on opposite sides of the earth at the same time but it is too sensible for scientists who prefer to teach that senseless gravitation produces the opposite high tides with an explanation that is so foolish that it is downright silly but willingly accepted as the truth by lying astronomers.

Revolving planets and moons given perpetual motions by a wonderful Intelligence is too reasonable for lying scientists who prefer to teach that perpetual motions of the planets and the moons are caused by senseless gravitation against an opposing centrifugal force which can not occur outside the air of the earth.

The constant changing attraction of gravitation if true and not wholly false, would produce only a wobbly Universe. The solution of gravitation was stumbled on, that scientists, Educators and Astronomers were believers and followers without reasoning faculties hence they can not escape from being blind followers and teachers of lying and senseless gravitation without being able to give a reason for their senseless lying.

This solution of gravitation seems to be true for it is impossible for any one to give a reason for teaching gravitation which is entirely void of reason but only dumb lying showing the impossibility of getting the truth about gravitation from dumb lying astronomers.

Universal gravitation as taught by lying astronomers is the acme of stupidity in Lying.

It is not likely that you can comprehend the Truth Sense and Beauty in the Universe on account of your scientific education.

Only a scientific mind with lack of common sense will teach senseless gravitation in which no one can find a word of truth or sense.

The Attraction of Gravitation is wholly imaginary hence a great mystery to the lying scientists for their inability to find truth or sense besides no power for action.

Yours truly
T.P. Stanley

Address: Moses M. Ashley
47 Crescent Street,
Long Island City, N.Y.
c/o Wadley.

Ladies and Gentlemen:-

If we believe in Brotherhoodism it is our enlightened duty to rally to a good cause, regardless of who is agitating such cause. This writer honestly regrets that our brand of civilization makes money so absolutely necessary, to the carrying out of enlightenment programs.

Therefore, this is to inquire if you are inclined to contribute to this cause? I have in mind the starting of a publication edited by me for the social uplift of this and future generations. I wish to put my reasonings in writing. Being at the crossroads of this brazen age, I think it the duty of any citizen to encourage that which is good. I have said enough for you to see my bent.

If you are uninclined to help this stranger to fight such sweeping battle against Infidelism and Atheism and Agnosticism, please recommend this program to a trusted friend of yours. I'll thank you so much. It may would be a good thing to help me in a disguised way to save yourself future annoyance. Find a way to do it. It is your duty to do it, or it is your friends duty to do it.

No, I am not a Clergyman.

Please show this to your spiritual adviser.

Yours sincerely,

Moses M. Ashley

Moses M. Ashley.

CAT. # 0023

LETTER FROM MOSES M. ASHLEY
TO
THE OBSERVERS AT MOUNT WILSON

CAT. # 0023

Address: Moses M. Ashley
47 Crescent Street
Long Island City, N.Y.
c/o Wadley.

Ladies and Gentlemen:-

If we believe in Brotherhoodism it is our enlightened duty to rally to a good cause, regardless of who is agitating such cause. This writer honestly regrets that our brand of civilization makes money so absolutely necessary, to the carrying out of enlightenment programs.

Therefore, this is to inquire if you are inclined to contribute to this cause? I have in mind the starting of a publication edited by me for the social uplift of this and future generations. I wish to put my reasonings in writing. Being at the crossroads of this brazen age, I think it the duty of any citizen to encourage that which is good. I have said enough for you to see my bent.

If you are uninclined to help this stranger to fight such sweeping battle against Infidelism and Atheism and Agnosticism, please recommend this program to a trusted friend of yours. I'll thank you so much. It is your duty to do it, or it is your friends duty to do it.

No, I am not a Clergyman.

Please show this to your spiritual adviser.

Yours sincerely,

Moses M. Ashley

CAT. # 0024

Letter from Edward
to
The Scientific Community thru
the Mount Wilson Observatory

CAT. # 0024

1

To the scientific world Thru the Mt. Wilson Observatory
Mt. Wilson Calif.

To whom it may concern:

This is to certify, That I have found the Key To all Existance. And all I ask of any one Is for them to read What I am about to say. Because it is not my purpose to tell What you already know. And consequently the proof Shall follow and establish My work to make it law.

For the key to all existance Is the key to the Law By which all things Come into existance and therefore my word Is the key to that law to be verified by proof Listen therefore to what I say As follows:

2

The Moon Is practically all Water frozen or Ice It was formed By water evaporating From the earth Which arose and gathered Between the Earth and Sun It is hollow Like a pumpkin The inside is composed of that part of the air known as Nitrogen And very very cold Consequently its water is frozen.

If the crust of the moon Was removed, it would be a Sun bright enough To destroy the earth. There is no life upon the moon, but Without the moon There would be no life upon the earth.

3

There is no land upon the moon Except what has been made By falling (debris) from the sky But without the moon There would be no land Upon the earth, and consequently all life Would be in the water. and Therefore we say that there is no life Upon the Moon by reason of atmospheric conditions. But There is an abundance of life In the Moon or in the waters Of the Northern half.

There is no rain there The light is always the same and the temperature has not changed One degree in a million years The water is H20. The Ice is as pure as snow and consequently the air is N40 and To make a long story short. That is a physical paradise.

4

Upon the (Southern) half however The sun has never shown, Except (thru) the reflections of The Earth and stars Which is approximately Five times greater Than the light of the moon Upon the earth. Where the days and nights are equal. Fourteen days of night and Fourteen days In the light of the earth. And not a very (—) place For life of any kind, and First to be explored by man.

Which will be done in time. These facts I will prove. In part, thru the Moving Picture World, by speaking a few words to them which will enable them to make a moving picture of the Moon in which it will appear to fall Upon the earth, and apparently come Within the reach of human hands.

5

The Earth, upon which we live Was created between the sun and the worlds center of gravity and fanned by the burning of Oxygen in the center (—) and made by the remains of life and the mixing of the elements From above and from below. And it lies in the (—) of Equal attentions and resistance Between the center of the Earth and the lines of Equal attractions and resistance above the earth. And consequently it is a hollow shell filled with Nitrogen.

Even as Nitrogen exists above the (earth) and the attraction and resistance From the center of the earth Is equal to that above the earth and an object placed below the crust of the earth, would fall upwards Even as it would fall downward If placed above the earth and with the same rapidity.

6

(Escaping) that the one (force) below would (loose) in velocity And the one from above would grow in velocity And so it is with the (line) of Equal attraction and Resistance above the earth. Where a common ball would flote like a feather and if two balls were liberated together, one above and one below The one above would go away From the earth. With the same increased velocity as the one below would gain In returning to the earth.

The Moon rests, or floats In that line of equal attraction and resistance - The Corona of the Earth and it cannot fall to the earth By reason of the earths resistance and it cannot fall from the earth By reason of the earths attraction and consequently it moves and floats In its orbit around the Earth.

7

If the crust of the earth was removed, a Sun would be revealed with such powerful Resistance, that it would Be cast out of the Solar System If the crust of the earth was doubled in weight It would sink Toward the sun One half the distance It is from it to day. And therefore by this Information, I will prove The thickness of the crust of the earth.

By the worlds great mathematitions, which will also determine the composition of all the (solar) planets and even that of the sun itself. First however I will tell them Their composition, that by (These) figures, they may prove the veracity of my (word).

8

The Moon moves in its orbit Of the earths resistance and the crust of the earth Moves in its orbit of resistance which in earths case is From the center of the earth and therefore it is not necessary for the earth To have any particular Thickness. And in fact there is no land at all covering the North Pole And if a common (ball) was (lessened) of the Earths Center of gravity, between

Them and the North Pole It would leave with such velocity that it would come To the surface of the water of the North Pole, providing The Ice was cleared away when it would have arrived and then it would sink and rest above the lower edge of the Earths Crust

9

and that is the substance of that universal Law which is given to all solar planets thru the sun. They were all created by him and (thus) they maintain his law and that same law came from the center of universal gravity (sh-) fourth In many generations-Many Suns in the center of as many systems. And therefore a careful study of our own Solar System means the comprehention

(To a certain extent) of the vastness of the universe. But at the present time, a thorough understanding is too strong For the human mind. Suffice it to say however That these few words will pave the way that leads to all understanding which lies within my hands.

10

In the spring of 1904, at the age of 29 I was admitted to the study of that Law and for thirteen years I carefully followed That voice which called me Son of Man. And since the month of June 1914 I have written many chapters which As yet are held in vain by Those to who I wrote. Therefore I have turned to those of scientific Research to give to them what I have by writing Ten chapters. and whenever I am recognized There will I abide To draw all men to me.

Know you therefore all of you That seek the truth in any capacity that I stand upon the foundation of the law from whence all things sprang And that I am able to deliver you From all misunderstanding. and In that capacity, I speak, not the Words of man but of Him That made All Things. I thank you Very Kindly.

12

Etholeum - The base of all existance - it is One with Electricity and There is no place where It does not exist. It is the conduit of The Light between all of the planets and thru the telephone and the radio and without it There would be no Earth Because there would be No sound. to be transferred between

The planets and Life upon the earth Etholeum, therefore is the Mother of all planetary existance and In her all things are begotten To be concieved and born of her (Providing they have not already Been born) the last of which Is human understanding.

13

Electricity the spirit of all existance. Is one with Etholeum and the positive extreme of all things made between it and Etholeum. And therefore the universe Is charged with Electricity By the center of universal gravity Thru all the (planets) Even as a battery is charged From the elements Than a (—) and the planets are (strange —)

By which they move and have their (—) In their regular orbits and therefore there is places in the universe where Some things do not (exist) but there is no place where these two extremes do not exist.

14

Nitrogen the first Form of creation between Electricity and Etholeum, (In which the light (opposed) In the center of universal gravity and Therefore it sprang to all Planets when then centers of gravity was formed.) Is that element that envelops The earth and the center that lies beneath the crust of the earth. Pure

Nitrogen beneath and Oxygen and nitrogen above the crust of the earth.

Mixed to the consistency of Life above and death below, the crust of the earth And so it is within and upon all of the planets of heaven, (—) that hydrogen exists In sufficient quantities upon Them, there is Life, but the center of them all is Pure Nitrogen.

15

Oxygen that element of the Air and water that envelopes the earth. In all the planets was created by Electricity in the midst of Nitrogen. and to day it is created By the light of the sun upon the surface of the water Between the air and the water. And so (perfect) is this (—) of creation that oxygen is always the same in the air and in the water.

Regardless of how much may be consumed by fire. and in fact the water and the air could not be without oxygen For in their primitive forms they were Hydrogen and Nitrogen and it took oxygen to make The water and the air, which was and is by the rays of the sun.

16

Hydrogen, was created by Electricity between Nitrogen and Oxygen and the three forms the Trinity of Life Even as Electricity, Nitrogen and Etholeum form the trinity of all planetary existance. Electricity the (passtime p) thru Nitrogen the passtime Entrance (—) Hydrogen between Nitrogen and Oxygen and these (—) forms the air and the water with the surface of the earth.and that of the water between which is the trinity of the worlds existance. By the gathering of the water below and above to form the firmament which in the beginning God called Heaven, and wherein we live.

17

And Life was created in the water Before the earth was formed And there it grew in great abundance Between the Air and the water and each atmospheric condition Brought forth new forms of Life (Until) the land appeared to support another form of life which came in the air (above) the same as it came in the water below the firmament. The (end) of what was found in the electrical radiance of the sun and

(Thereby) planted in the water and upon the land and there it grew and passed away according to atmospheric conditions thru out the ages that led onward toward the earths (perfection) until the present times.

18

Man was created at the end of the day of imperfect atmospheric conditions and his birth created a (Kingdom) to rule the earth and How man endowed with the possibility of Thought Knowledge and Understanding which is the (—) of human perfections When the three are one. But until the present time the truth has been withheld By reason of human ignorance.

and the inability to grasp it when it appeared in sufficient strength to reveal the creation of the earth and universe and the human mind never would Have been sufficiently strong, had it not Have been for astronomical and scientific research, and therefore I hereby commend you all to them.

19

The Father, the Son and the Holy Ghost In human creation means the same as Electricity The Sun and Etholeum In Solar Creation With the throne in the Sun The Center of the Solar System Even as the Son of Man Is the (Throne) of Human (Existance) In the midst of Men Then to overcome all Mankind, even as the Solar System is (ruled) by the sun God, the Father, in all forms of Creation Is that same, (power) Father or God, that First came into the center of Universal Gravity

The Sun or Son of God Is His throne whether physical or mental and due to prevail upon the earth Even as the Sun prevails in the universe the Holy Ghost is the possibility of Creation, the negative (power) to receive the passive spirit, in order that the created may arise between the Holy and the Unholy.

20

Thought, Knowledge and understanding Is also the (—) Those ancient words. Which is the trinity of that Kingdom To rule the Earth and therefore the same condition Exist to day among men, that did exist in the beginning Before the sun was born. We have the two (entrances) But Knowledge has not arisen between Thought therefore in every human mind is simply another word meaning God within us

Knowledge means the return of The son of man and understanding means that he will draw all men to him By the proof he holds within his hands Because the Father, the Son, and the Holy Ghost, or Spirit is to every human mind the same, as Thought Knowledge and understanding which is divine.

21

Government, of by and for the people Is another trinity to (—) the forward stride of civilization By making laws to be obeyed by some and disobeyed by others. For the (—) of some and the destruction of (others) and Founded upon the propositions of Equal Liberty and Justice for all. Government of all people, is of those that make the Law By those that obey it.

And for those that (brake) it. And when the Fathers are one with the Sons, by obeying the laws they make, the law will be inforced in the far extreme and the human race will be one Indivisible people with Liberty and Justice for all.

22

The Election of Law Officials by the people for the purpose of Inforcing the Law Is the transforming of Government From the hands of the people to (—) of a few and therefore If the officials of the Law are obedient to it, To begin with, the Law is inforced by a two thirds Majority, (—) the election should terminate, In the honesty of Alfred E. Smith who made this Immortal Statement: "Hoover is the not the president of the Republican Party But the president of the United States" Like the Father and the Son he (tries)

To these convictions and the Law to begin with, Is Inforced by a two thirds majority And it shall prevail In the far extreme Until Crime shall vanish from all the Earth Never to return of any great consequence and never in the majority Beyond the power of the law to (—) it.

23

And now I must conclude My words to you By saying that the constitution and its Eighteen Amendments shall never fall But by these four last chapters 19, 20, 21 & 22. It shall be transformed From a mighty oak tree to a royal palm. Because these four chapters are four more amendments To it, to grow thereon. Which I hereby Trust to you of the Astronomical and the scientific world.

Because you have used the talents of your minds (—) and Have not hid them in the and consequently you are (worthy) to receive my words. In witness where of I set my hand and affix the (seal) of (—) By signing my name - Edward.

The End

LETTER FROM FREDERICK K. DENTWILLER
TO
THE OBSERVERS AT MOUNT WILSON

CAT. # 0025

Frederick K. Detwiller
Carnegie Hall
New York

Description
of Paintings of
The Eclipse of January 24, 1925

—oOo—

In my three paintings I desire to show the story of the event as it relates to our life and bring a romance to astronomy in terms that children and the lay mind will understand and hold the event by a symbol of art, which is depicted in three stages, or a Tryptic.

1st Painting - The Processional, or Bridal March, as the Moon comes to wed the Sun, attended by Venus, Jupiter and Mercury, etc. Stage - Just before Totality.
Size 31 x 44.

2nd Painting - Totality - The Marriage attended by the Heavenly Hosts.
Stage - Totality, etc.
Size 60 x 44

3rd Painting - The Ring, or Recessional - the symbol of the Act as the Moon now goes away, etc.
Stage - Just after Totality.
Size 40 x 44

April 11, 1929.

Frederick K. Detwiller
Carnegie Hall
New York

The Marriage of the Sun and Moon
at Haverstraw, N. Y.
on January 24, 1925.

—oOo—

The Eclipse

Observations were made at Haverstraw, N. Y. on the morning of January 24, 1925, from the Look Out at the end of Main Street.

Looking south towards New York City as the moon shadows appeared, I noticed it soon grew dusk as a wierd light illuminated the landscape to the East bank of the Hudson. Towards the West things grew dark, people very black, and I could not register the color as it was unearthly, and at 8 minutes before Totality I went for one minute into a dark room in the U.S. Hotel to refresh my mind and eyes and returned to the Look Out.

My viewpoint was to record beauty as seen by the corporeal eye of the artist - the values and color of the phenomena. No technical instruments were used only my regular outfit and sketch book and a 5-B Pencil for quick notes.

We were now in the creeping shadow of the Moon, greenish in color and growing dark in values - uncanny.

Suddenly the Moon covered the Sun in Totality. Startled I jumped. I could only in awe gaze at the Corona and stood transfixed by a pearly light, beautiful beyond celestial description with a rim or edge moving in perfect Harmonic Motion. The

Moon like a disk was black and dark green in the center. Hence the green shadow generally recorded.

My mind now spoke! Get the landscape, so I record the impressions as follows: that the law of values had been completely reversed, i.e., the Moon was the darkest value or blackest of all things. The law of values as to distance of object vanished because that which was thousands of miles away, or farthest removed, was the darkest.

So I continued to scan the landscape - down the Hudson beyond the Hook Mountains and to the center of Vanishing Point. The River was frozen in an Ice Flow with wind-swept ice and snow, reflecting colors of rare hue and streaked in wonderful designs, which mirrored the lights of the Corona on the glassy ice and in the snow-covered areas. The colors in shadow were dark green, dark violet and dark orange, in all variations; in the light, blues, reds, and yellows, and white, in all values. The ice flickered like rare, unknown jewels. The Hook Mountains stood out in a sharp silhouette against a dawn sky together with the distant mountains on the East Bank. Against the silhouette on the Hook Mountains the trees appeared in sharp patterns and snow showed distinctly in shadow.

In reality we were in a Great Shadow looking at the distance all bathed in a cold light reflected from the Corona and Moon.

The phenomena in shadow was not like a night in moon- light because here everything was in sharp line and edge, whereas in moonlight the edges are soft, lost and chewed off.

In the foreground there were boats, houses and tanks. The trees stood in shadows like ghosts in dark purple against the ice and snow.

The sky above the Hook Mountains was in its effects a most unusual colorful dawn, - high lights on the filmy clouds - some cold, some warm.

The sky directly above the distant Hudson was warm towards the South, and back of the Hook Mountains, in the West, cold, queer greenish and Cerulean blue.

The sky at the Zenith was dark blue and violet. The three planets, I noted the last seconds before the flash of light stood out in a clear aureolean yellow against a deep Sky, were Venus, Mercury and Jupiter.

I believe for color observation Haverstraw was the ideal place that zero morning because we had the shadow of Totality and Dawn. This is possible not in the center of the shadow band but is better towards the edge.

During Totality it is the pearly light of the Corona which gives comfort and hope to Earth. Should that light go out, all would be lost.

While the papers of the country did a great service, the scientific aspect was always brought to the fore and the beauty and rarity of the occasion, which should have had a place, was not featured prominently.

To sum up my experience,- the Hudson and Hook Mountains brought one for a short space of time to the Arctic Circle not unlike Alaska. The snow like glass, with the wonderful yellows, purples, blues and greens on the River, made a deep impression.

I am grateful to have seen so wonderful a spectacle and in such beautiful surroundings and in so rare a place by the River's edge, and wish to remark, "many are called but few are chosen."

Frederick K. Detwiller

Haverstraw, N. Y.
January 24, 1925.

Notes: Added January 26, 1925.

I had two guests on each side during the event - Miss Van Orden, the librarian at Haverstraw, remarked, "Red Flames directly over the Moon."

Mr. William Lyons, distinguished artist, whom I brought up from New York, made sketches immediately afterwards in the Impressionistic manner in oils, feeling its occult and spiritual significance.

Strange remarks:

One man said of it, "A picture no Artist can paint! But time will tell. Who knows."

The Mayor of New York City, John F. Hylan, quoted from the New York World - Sunday, January 25, 1925:

"The Sun may be eclipsed but New York never."

The Mayor of a city in our Republic is elected by the People and is supposed to reflect the attitude of the majority. This episode is given so you will note how much we have advanced in things spiritual since antiquity?

X- Ray Drawing---

Explained in Eze. 1st & 10 Chptrs
Also Mentioned Rev. 4th Chpter***

ORION NEBULA
***** Dwelling Place of The Gods*******
Understood by man in 600 B. C. as represented by Symbols of Face of Man, Lion, Ox, Cherub & Eagle--- said forms and resemblences found on Photographic plate.

By W. Chas. Lamb

Five Faces, 4 Wheels, 2 Cherubims, 8 of the 16 Wings, 2 Wings Covering Body(dark portions, Corlors of Rainbow, Terrible Crystal stretched forth & etc--- very interesting to locate 33 points of identity-- with that of scripture.

FACE OF CHERUB

OX

LION

CRYSTAL

SERAPHIM

OX

WINGS

TO TRAPESIUM AND 4 WHEELS
AND 2 LITTLE STARS CALLED CHERUBS
RESIDENT OF THE GODS ESEKIEL 10: 9 VRS

MAN FACE

EAGLE

DARK PORTION
...ing THEIR BODIES

TURN DRAWING AROUND

CAT. # 0025

LETTER FROM W. CHAS. LAMB
TO
DR.WALTER S. ADAMS
CAT. # 0026

X- Ray Drawing---
Explained in Exe. 1st and 10 Chptrs
Also Mentioned Rev. 4th Chpter***

ORION NEBULA
***** Dwelling Place Of The Gods *****
Understood By Man In 600 B. C. As Represented By Symbols of Man, Lion, Ox Cherub & Eagle--- said forms and resemblences found of Photographic plate.

By W. Chas. Lamb

Five Faces, 4 Wheels, 2 Cherubims, 8 of the 18 Wings, 2 Wings Covering Body (dark portions, Corlors of Rainbow, Terrible Crystal stretched forth & etc--- very interesting to locate 33 points of identity--- with that of scripture.

Note---- Cherubim & Serupim are two of the most powerful orders of Angels mentioned in scripture--- undoubtidly their position is on either side of the Throne of God.

3 Personalities constitute the God head

Lucifer-- Now Called Satin--- was once a covering Cherub, and his position was undoubtidly near the 3rd person of the God-Head-- according to scripture-- He fell from this position 6,000 years ago-- and deceived Eve-- Thus two Cherubims and 4 Wheels.

Saith Scripture-- "God Establisheth His Throne in the Heavens, and placeth a cloud upon it." Job.

Astronomy has revealed the appearance of a Horse like dark Nebula in the vacitiny of the Orion Nebula--- What is it?

Was primitive man for ages mistaken in their conception of the Gods, when they cut these creatures in stone, for posterity?

We might thank Ezekiel for his version on the Proposition of the Gods.

W. C. L.

P.S. Also it is wonderful that the Mount Wilson & other observatories are taking such wonderful photographs of this marvelous and stupendious phenemonon.

----- Truly a great age-----

President Coolidge couldnt be the entire U. S. and Neither could the Gods be the entire Orion Nebula.
Either have to have a Map to work on.

W. C. L.

(Important)
(Strays from the ether revealed to man 600 B. C. what we are just now beginning to find out through Modern Astronomical photography)

Hubbell, Nebr. 7/26/28

Dr. Walter S. Adams
Mount Wilson Observatory,
Pasadena, California.

Dear Mr. Adams:— I thank you for your letter of the 23rd, in answer to by letter of recent date pretaining to the resemblance and forms found in the Orion Nebula and note that you cannot offer any suggestions regarding my views.

These forms are only an inkling — just a trace — a readily intelligible and even forceable mode of representation of "That Consuming Fire" the Almighty, understood by symbols, and traceable in early Chaldean Art, to-day dug up in the ruins of Ninevah and Babylon, such as winged, human headed Bulls, winged Lions, heads of an Eagle or Vulture, a curved beak, half open, disclosing a narrow pointed tongue & etc. & etc. all elements (including the square vessel and the fire cones) simbolizing fire, with that of the qualities of the homes of the Gods.

The Nebula its-self is not God, any more than the 7 stars, other than representing the 7 Spirits of God. Undoubtidly they point us to the region in the heavens, where dwell the Gods, and these Symbols had awed and instructed races which existed on this planet centuries B. C.

In the light of Modern science and the intrepretation of the scriptures, I claim that these forms co-incide with the teachings of the latter — an act of the Devine Spirit in His effort to reveal Him-Self to man.

"GOD DWELLETH BETWEEN THE CHERUBIMS", and Ezekiel 10:9 verse, reveals the secret.

"One wheel (disk by on Cherub, and another wheel by another Cherub (Father & Son) and the appearance of the wheels was as the color of the beryl stone."

Note — The five faces (man, lion, ox, eagle, & cherub) are forms and resemblances found only in the Orion Nebula, the 4 wheels are found in the TRAPEZIUM, and you will find as explained in scripture 2 little stars besides two of the larger. Found on the photographic plate, to be as recorded by Ezekile in 600 B. C.

The stars in The Trapezium therefore would represent to the Human mind the positions or Thrones of the 4 important objects mentioned in scripture — THRONE OF THE FATHER, THE SON, THE HOLY SPIRIT, & THE HOLY CITY — this is my position, proved by the photograph, & the Scriptures, - thus the dwelling place of the ineffible glorious and incarnate God, — not that the Orion Nebula is God — but the living creations surround His Great & Glorious Personage — what ever that may be.

The word - Wings - translated from the Chaldeian into the english would signify — the extremity of a country — Eyes — guided by intelligence — FIRMAMENT — a solid expanse, or platform, elevated floor & etc. — Terrible Crystal — dazzling and Brilliant, fearful to behold — all Living Creations — Temples not made with hands —- and the prophet saw the appearance of a man upon the throne. That Man is undoubtedly Jesus Christ, the Wisdome of these LIVING CREATIONS — who was God in the flesh — although God the Father is Spirit — and undoubtidly resides in or on the Largest Disk in the Trapezium found in said Nebula. — thus according to scripture — the two Cherubims occupying (highest order of Angels) — standing on either side of God and His Son.

Pardon this intrusion — but if you will study the scriptures and the photographs, you will find probly more than 33 points of identity — proving the dwelling place of Gods — in The Great Nebula of Orion.

Very Truly Yours
W. Chas, Lamb

Important--- (Just now while scientists are digging up evidence pretaining to the Gods of the Ancients--- ON THE OTHER SIDE OF THE WORLD-- at least their is one person on earth that knows that astronomers are finding the actual evidence of the True Gods in the Heavens-- through modern astronomical appliances-- also NOW-- (thats me!)

Hubbell, Thayer Co, Nebr. 7/29/28.

Dr. Walter. S. Adams,
Mount Wilson Observatory,
Pasadena, Cal.

Dear Sir:--- Just another minute of your time-- please read the enclosed-- news--- "Strange Discoveries in the Tombs of Ancient Ur," where I have penned #'s 1, 2, 3, 4, 5, 6, 7, 8, showing what scientists are finding out about the Chaldees conception of GOD and their mode of worship no higher than the Moon-God. Their symbolism pointed to the Bull, as representative of strength-- but the Lion hadn't yet pointed to "The Lion of The Tribe of Judea--- God incarnated in the flesh-- couldn't see how it could be done!

Thus this ancient people tried to embody their conception of the wisdom, power, and ambiquity of the Supreme Power or Being that rules the Universe-- their five senses took them to the Sun & Moon Gods-- didn't know that some day their would be telescope, and the art of photography-- that would actually bring out these resemblences-- in a Nebula.

600 B. C. or thereabouts, Mr. Ezekiel, through some mysterious means, had a glimpse of the true Gods in the heavens-- not the Moon or Sun God-- and does his best in discribeing the much (what ever that is) that goes to make up-- the residence of the Gods & The God.

I believe this proposition of finding resemblences in the Orion Nebula of the God or Gods as recorded by Ezekiel, and found exactly as discribed on the photographic plate-- is quite a feat-- whether we realize it or not. We have approached very close to a property - that we might consider dangerious-- "NO MAN CAN SEE GOD AND LIVE." (thats Scripture)

Our nerves wouldn't stand the "salt" so to speak-- but Mr. Ezekiel, being a Chaldeen, got by in some manner, not thoroughly understood by our modern scientists. How he could stand the luster and effulgence of Glories of, "WONDERFUL, COUNSELLOR, the MIGHTY GOD,-- shows his physical make-up must have been made out of some other kind of atoms, other than Tunney or Henneys (prize-fighters), of our age, said to be able to with-stand much punishment.

How much longer will the world have to wait, untill this proposition in a way is settled, as to what is, where, and who is-- GOD? It would seem to me that said questions can be proven scientifficaly, and in accordance with the scripture-- and in closing-- I am going to state-- that I believe-- that this is a good theme to hit upon-- and that we are just now entering upon the FOOL AGE-- when this class is very apt to confound the WISE!

Just put your-self back 5,000 or 5 Million years -- and compare then with now-- all they needed was a little FAITH-- but to-day we can put all the faith in the world inside of a Mustard Seed-- because I have never heard of a Mountain being moved-- by FAITH!

With kindest regards & best wishes, and hope I haven't bothered you, I am

Very Truly Yours,

W. Chas. Lamb,

P. S. Just ran onto this article to-day, and thought it no harm to mail it to you.

P. S. "Oh Fools, and slow of heart to believe all that the scriptures hath spoken. Had Not Christ ought have suffered these things, and then to have entered into His Glory ." (sayings of Christ)

(Strange they are digging it up over their, and finding IT over here?)

CAT. # 0026

LETTER FROM W. CHAS. LAMB

TO

DR. WALTER S. ADAMS

CAT. # 0027

Important — (Just now while scientists are digging up evidence pretaining to the Gods of the Ancients — ON THE OTHER SIDE OF THE WORLD — at least their is one person on earth that knows that astronomers are finding the actual evidence of the True Gods in the Heavens — through modern appliances — also NOW— (that's me!)

Hubbell,
Thayer Co, Nebr.
7/29/28.

Dr. Walter. S. Adams,
Mount Wilson Observatory
Pasadena, Cal.

Dear Sir: — Just another minute of your time — please read the enclosed — news — "Strange Discoveries in the Tombs of Ancient Ur." where I have penned #'s 1,2,3,4,5,6,7,8 showing what scientists are finding out about the Chaldeian conception of God and their mode of worship no higher than the Moon God. Their symbolism pointed to the Bull, as representative of strength — but the Lion hadn't yet pointed to "The Lion of the Tribe of Judea God incarnated in the flesh — couldn't see how it could be done!

Thus this ancient people tried to embody thier conception of the wisdom, power, and ambiquity of the Supreme Power or Being that rules the Universe — their five senses took them to the Sun & Moon Gods — didn't know that some day there would be telescope, and the art of photography — that would actually bring out these resemblances — in Nebula.

600 B. C. or thereabouts, Mr. Ezekiel, through some mysterious means, had a glimpse of the true Gods in the heavens — not the Moon or Sun God — and does his best in discribing the much (what ever that is) that goes to make up — the residence of the Gods and The God.

I believe this proposition of finding resemblances in the Orion Nebula of the God or Gods as recorded by Ezekiel, and found exactly as discribed on the photographic plate — is quite a feat — whether we realize it or not. We have approached very close to a property that we might consider dangerious — "NO MAN CAN SEE GOD AND LIVE" (thats Scripture)

Our nerves wouldn't stand the "gaff" so to speak — but Mr. Ezekiel, being a Chaldean, got by in some manner, not thoroughly understood by our modern scientists. How he could stand the luster and efflulgence of Glories of the "WONDERFUL, COUNSELLER, the MIGHTY GOD, —shows his physical make-up must have been made out of some other kind of atoms, other than Tunney or Henney (prize-fighters), of our age, said to be able to with-stand much punishment.

How much longer will the world have to wait, until this proposition in a way is settled, as to what is, where, and who is — GOD? It would seem to me that said questions can be proven scientifically, and in accordance with the scripture — and in closing — I am going to state — that I believe — that this is a good theme to hit upon— and that we are just now entering upon the FOOL AGE — when this class is very apt to confound the WISE!

Just put your-self back 5,000 or 5 million years — and compare then with now — all they needed was a little FAITH — but to-day we can put all the faith in the world inside of a Mustard Seed — because I have never heard of a Mountain being moved — by FAITH!

With kindest regards & best wishes, and hope I haven't bothered you, I am

Very Truly Yours,
W. Chas Lamb,

P.S. Just ran onto this article to-day and though it no harm to mail it to you

P.S. "Oh Fools, and slow of heart to believe all that the scriptures hath spoken. Had Not Christ ought have suffered these things, and then to have entered in His Glory." (sayings of Christ)

(Strange they are digging it up over there, and finding IT over here?)

EPPLEY HOTELS CO.
OPERATING

The Lincoln
AN EPPLEY HOTEL

Lincoln, Neb. Dec. 21,31.

Dr. W. S. Adams,
c/o Lick Observatory,
P sadena, Cal.

Dear Dr. Adams:--

Here is the method of telling the age of our Solar System, and evry other system in the Universe.

Our Solar system originated in the Nebula of Orion. It is traveling in a curved line for myrids of milleniums. Add the distance in light years, and make allowance for the curve vature of our solar systems orbit. Devide this result by the speed of our solar system through space, and it will equal about 36,000,000 approximately. Evry system in universe measured on this plan, and exact age determined, nebulas & etc.

******** Proof *******

The Orion Nebula is that consumeing fire called God, or in which the Devine Being or Creator resides. Visions of God were seen in the Nebula about 600 B. C. Ezekiel. I: I, Five faces were seen, face of man,Lion, Ox, Eagle and Cherub. FOUR WHEELS in trapezium, and twp Cherubims, a total of 6 Suns or stars. Also I6 wings, or wisps of Nebulas matter.

Read Ezekiel. I:Io. Io: I4; Rev. 4th chapter.

When Our System was started out36,000,000 years ago, God said let the sun rule the day, and let it be for days, weeks seasons and years & etc.

Thus our Solar System started on its journey.

CAT. # 0027

LETTER FROM W. CHAS. LAMB
TO
DR. WALTER S. ADAMS

CAT. # 0028

Dec 21, 31
Dr. W. S. Adams,
c/o Lick Observatory,
Pasadena, Cal.

Dear Dr. Adams:— Here is the method of telling the age of our Solar System, and every other system in the Universe.

Our Solar system originated in the Nebula of Orion. It is traveling in a curved line for myriads of millenniums. Add the distance in light years, and make allowances for the curvature of our solar systems orbit. Divide this result by the speed of our solar system through space, and it will equal about 36,000,000 approximately. Every system in universe measured on this plan, and exact age determined, nebulas & etc.

*******Proof*******

The Orion Nebula is that consuming fire called God, or in which the Divine Being or Creator resides. Visions of God we seen in the Nebula about 600 B. C. Ezekiel. 1:1, Five faces we seen, face of man, Lion, Ox Eagle and Cherub. FOUR WHEELS in trapezium, and two Cherubims, a total of 6 Suns or stars. Also 16 wings or wisps of Nebulous matter

Read Ezekiel. 1:10. 10:14; Rev. 4th chapter

When Our System was started out 36,000,000 years ago God said let the sun rule the day, and let it be for days, week seasons and years & etc.

Thus our Solar System started on its journey.

Our solar system speeds through space at the rate of about 400,000,000 miles per year. 6000 years ago, it was not where it is to-day. 36,000,000 years ago it was Created at the Beginning. Gen. 1:1 verse & etc.

You can get a lot of kick figuring out how Old Betelgeuse or Arcturus is. Or some of the stars out in space up to a hundred Million - Light years away from the Orion Nebula

God never went to each of the 64 Trillion Solar systems and created them, any more than Henry Ford. would move his factory to the South Pole to build a Ford there.

All Creation has taken ages and was brought into existence in the Vicinity of the Orion Nebula, unless it can be proven that their are other Gods.

We are warned to not have Any Other Gods before Him.

And so I am asking you to please try this plan out, look up the references, and symbols of Deity, as mentioned in scripture, and I believe that you will find the method of finding the age of any solar system is very nearly correct. Please tell your other co-workers about this method,

Very Sincerely Yours,
W. Chas. Lamb,

P.S. You will see the face of Man, Lion, Ox, Eagle, and Cherub in a good photo of Orion Nebula. Also 6 suns in trapezium.

LETTER FROM R. M. WIDNEY, ATTY.
TO
DR. FRED E. WRIGHT

CAT. # 0030

R. M. Widney, Atty
714 S. Hill St. Los Angeles, Calif
Aug. 9, 1928

Dr. Fred E. Wright
Hotel Pasadena, Cal.

Dear Sir:

Yours of the 7th inst., replying to my letter of the 6th inst., relating to revolution of the moon on its axis, at hand. Having carefully considered the same I think there is an error in your demonstration which will appear by the following illustration:

Fig. 1:

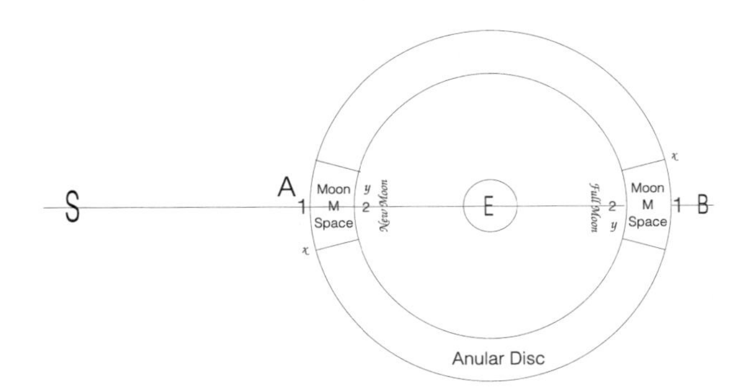

The Anular disk represents a part of the plane of the moons orbit around the earth. The Sections xy represent a moon space.

This moon space cannot revolve on its axis during a revolution of the disk around the earth. Sun moon and earth being in position A.M.E. the face (2) of the moon is toward the earth; face (1) is seen from the sun. This face 2 is always toward the earth,(E). Face (1) is always from the earth. -1-

Starting at position "A", (new moon); and, revolving the Anular disk, with the moon space as a part of it, half of a revolution to B, (full moon) the sun will shine on face (2), seen from both sun and earth, moon space xy has not revolved on its axis. That is impossible- being a fixed part of the Anular disk. From the sun an observer would thus see all faces of the moon space xy and might infer, as explanatory, that the moon space xy revolved on its axis. Whereas his observations would really be accounted for by the revolution of the moon space in the orbit around the earth. If the moon revolves on its axis once each orbital revolution, then at point B, a half axial revolution would be completed and face (1) would be toward the earth, and thus the earth observer would see all sides of the moon each revolution. Assuming that the moon has no axial revolution, it follows:

1. The same face of the moon is always toward the earth.
2. To an observer on the sun the moon would have an apparent axial revolution, mathematically corresponding to an orbital revolution of the moon around the earth.
3. He might think that the moon revolved on its axis to produce the actual phenomena; whereas, the actual phenomena as observed from the sun is due to the orbital revolution of the moon, and is not due to any axial revolution of the moon.

With reference to the earth the same face (2) of the moon alway being presented to the earth is the same in effect, as if a rigid rod radius held the moon in that fixed position preventing axial revolution.

The mathematical coincidence of an axial, and orbital revolution of the moon has never been accounted for.

If not too much trouble I will appreciate a reply from you.

Truly yours,
R. M. Widney
RMW:MEL

LETTER FROM E. L. EWELL
TO
DR. SETH B. NICHOLSON

CAT. # 0031

May 10, 1932

Seth B. Nicholson
Pasadena, Calif.

Dear Sir;-

You speak of the angular distance of the N. star from the north celestial pole. My question was to know the distances (not an angular distance) in statute miles. Does One degree 30 minutes mean an angle or about 90 statute miles between the north celestial pole & the north star, in the vacinity of the star,? That is the question.

You say sunshine is Radiated heat, from the sun 93 million miles. If that was true there could be no cold between the earth & sun, but it is cold every where 3 miles up. & in 93 million miles you have the heat reduced to only thermometer capacity. According to this the remaining heat to go farther, to warm these poor fellows on the superior planets would be short & cold long before reaching them, and just pity those scorched people on Venus & Mercury.

Your Question - can Vacuum be hot or cold?

Ans. Vacuum, where housed in small space, & surrounded by temporate degrees, is temporate, but the Vacuum of Universal Open Space is a different condition, Its temporature is cold, & 93 million miles of it would never in any length of time be penetrated nor heated by any intensity of heat radiated in all directions from any ball of heat.

Why don't you get busy trying to win my $25. reward,?

Very truely

F. L. Ewell

Letter from T.C. Bates to Dr. Wilton Humason

CAT. # 0032

Dr. Wilton Humason
Mount Wilson Observatory.

Dear Sir:

What can be the matter with our leading scientists and philosophers? Do they mean what they say? We think they are overlooking a great thing, the fact that human reason has its limits. We think they make the great mistake of trying to solve the insoluble in their attempt to reason outside the domain of human reason.

It is said that our astronomer, Mr. See, professes to be near the solution of the mechanical workings of gravitation, but he tries to advance his researches along this line by comparing the mechanical workings of gravitation to that of the corkscrew. Here he makes the mistake of trying to solve the insoluble. This problem is insoluble, at least, in our present state of knowledge, but we do not say that it will ever remain so, because we do not know that our knowledge will never advance far enough to give us a clue to the solution of this problem.

By comparison, we can tell what gravitation is not like but we cannot tell what it is like. To explain the workings of gravitation, by comparison, there must be an analogy between the things compared. Looking into this problem, we find that we can detect no analogy between the mechanical workings of gravitation and that of anything within our knowledge, therefore, in the present state of our knowledge, this is an insoluble problem. We cannot reason on it at all, in the way that Professor See reasons on it, or tries to reason.

The futile attempts that are being made and have been made in trying to reason outside the domain of human reason is, to me, surprising. We hold that all thoroughly trained minds should know when they are trying to reason outside the sphere of reason and should, therefore, cease their efforts and await further developements. Mr. See, as we find many other scientists do, including myself, but I am not a scientist, makes many mistakes within the domain of human reason. Now I do not profess to be a prodigy. The writer has never been inside a college building; never conversed with a man of science; but has been acquainted with a few profound though obscure and uneducated philosophers. So I hope you observatory fellows will bear with me in my criticisms and not give me the cutting answer that some of the men of science and scientific papers have done. The Scientific American says "Frankly, we do not think there is anything in your pretensions."

It is said that of a ray of light from the sun, scientists say so, only one threehundered thousanth of it strikes the earth. Professor See accounts for the variation of the sun's heat by meteors falling into the sun. Now to increase the temperature of the earth three degrees, the meteors would have to raise the 900,000 degrees. Such a downpour of meteors would wreck the solar system.

Again, astronomers say that, were Mercury to fall into the sun, she would produce enough heat to make up for that lost by radiation in seven years; but they do not make any allowance for the heat absorbed from the sun. For all we know, its fall would lower the temperature of the sun instead of raise it, but much more could be said on this subject, but space forbids.

Dr. Jeans says that the world is on its way to extenction without any chance to reverse, at least I read in the papers that he says this. This appears, to me, ridiculous. If the universe is infinite in time, and all indications appears to me that it is, it must have been becoming extinct from everlasting or in some past time it began to become extinct. Reason teaches us that it could not have been becoming extinct from everlasting and if it is eternal and commenced extinguishing itself at some past time, then an eternal past time elapsed before it began to extinguish itself; this is absurd, because reason teaches us that a condition or state that has lasted from everlasting will continue during the everlasting future. Mr. Jeans says that we are ignorant as to how the universe began. How does he know it began? He thinks there might be a materialistic interpretation of the pouring down of immense amounts of radiation into empty space. The idea of space is the idea of the absence of all things. ? by radiation pouring into empty space, where did the radiations come from?

In our youth, through ignorace, we regarded the scientists and the philosopher as something almost superhuman, but becoming old (77) and having more time to study, we have rid ourself of this delusion. Gathering together our past reflections and reinforcing them with our present ones, we could sit at this desk for six months in enumerating just such mistakes as given above, still, the scientific papers absolutely refuse to publish any of my writings.

But what about our man, Einstein? What does he mean? Is he just kidding us or is he crazy? Please understand that we do not hold that everything he says is foolish. We do not deny that light rays are bent by the sun's attraction, though this may be caused by refraction. We do not deny that he is a great mathematician, but we hold that the most of what he says about relativity is ridiculous and absolutely absurd. Einstein is a great puzzle to me, but the greatest puzzle is why scientists pay so much attention to his foolishness.

Einstein has a great deal to say about space and time. Here he displays remarkable shortcomings. The ideas of space and time are ideas that aid us in reasoning and arriving at conclusions. Being equally distant from the gate, the horse will run faster than you in order to go through the gate. Here the horse has just as correct ideas of time and space as the greatest philosopher. We never see a horse try to leap a across chasm twenty feet broad. He will not try to beat you to the gate if you are too close to it. The honey bee has just as correct idea of time and space as Albert Einstein. What Einstein and other would-be philosophers are trying to do is to improve on the horse's idea of time and space by regarding them as concrete existences, while the ideas of them are imaginary and perfect so far as they are intended to go in helping our reasoning capacity. All philosophers have been trying to reason outside the domain of reason in their attempts to improve on our definitions of space and time. Time is the idea of the interval between two events. This interval may be imaginary. If the universe is eternal in time, we may imagine an interval as existing between the eternal past and the present.

I hear that Einstein intends to visit your observatory in the near future. Hope he will, but he, with our help should be careful that he does not go off at a tangent, for, if a straight line is crooked, a curved line is straight. If a straight line returns upon itself a crooked line must go on forever. This is my philosophy. Now the line from Europe to the observatory is crooked and if he is not careful he will not arrive at the observatory. A New York professor was about to try Einsteins insulation-of-gravitation theory. He thought he could walk on the air, if insulated, but I wrote and told him to be careful, and not try it in himself but try it on a monkey, because should he try it on himself and succeed, instead of walking on the air, the pressure of the air would shoot him up at an accelerated speed, and, according to Newton, a body once put in motion and acted on by no force continues to move forward in a straight line forever; so he might never return. We have not heard from him, but we hope our letter reached him in time.

Well, brother, you may publish this letter or do what you please with it, but we would be glad, should you pass it around among your observatory fellows and finally answer and tell me what you think of it.

Most Respectfully Yours,

T.C.Bates,
little old twenty acre clodhopper
Mead, Wash.

Letter from H. R. Hildeman to the Observers at Mount Wilson

CAT. # 0034

Manawa, Wis.
Feb 14 '29

Mount Wilson Observatory
California

Gentlemen:

Why don't you use those powerful new rays, aiding vision but not killing, directed to the surface of a planet, lighting up an acre, more or less, and shifting about. Then, with your giant reflectors, read the story of Mars and also study cell-division (Nuclei) of the nebula and the diffused cells in the rings of Saturn.

With apology

H.R. Hildeman
(Farmer - scientist.)
Manawa, Wis.

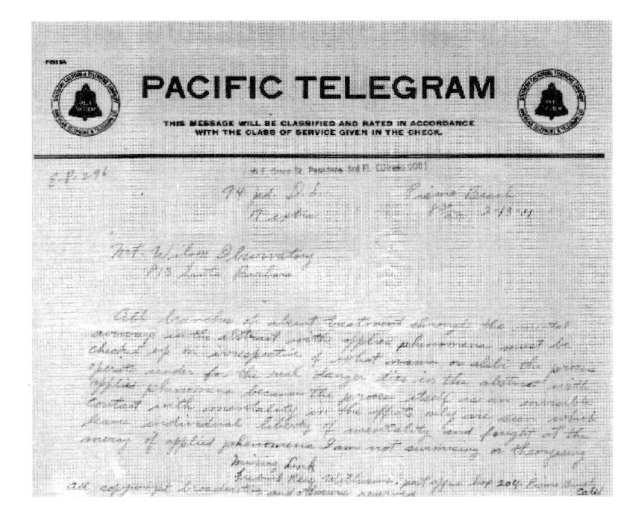

PACIFIC TELEGRAM

THIS MESSAGE WILL BE CLASSIFIED AND RATED IN ACCORDANCE WITH THE CLASS OF SERVICE GIVEN IN THE CHECK.

E-P-296

94 pd. D.L. Pismo Beach

17 extra 8 30 am 2-13-31

Mt. Wilson Observatory
813 Santa Barbara

All branches of absent treatment through the mental airway in the abstract with applied phenomena must be checked up on irrespective of what name or alibi the process operates under for the real danger lies in the abstract with applied phenomena because the process itself is an invisible contact with mentality and the effects only are seen which leave individual liberty of mentality and fought at the mercy of applied phenomena I am not surmising or theorizing

Missing Link

Frederick Rey Williams, post office box 204 Pismo Beach Calif

All copyright broadcasting and otherwise reserved

CAT. # 0035

TELEGRAM FROM FREDERICK REES WILLIAMS
TO
THE OBSERVERS AT MOUNT WILSON

CAT. # 0035

800 E. Green St. Pasadena 3rd Fl. COlrado 0001
E.P.296 Pismo Beach 94 pd. D.L.
8:30 am 2-13-31 17 extra

Mt. Wilson Observatory
813 Santa Barbara

All branches of absent treatment through the mental airways in the abstract with applied phenomena must be checked up on irrespective of what name or alibi the process operate under for the real danger lies in the abstract with applied phenomena because the process itself is an invisible contact with mentality in the effects only are seen which leaves individual liberty of mentality and fought at the mercy of applied phenomena I am not surmising or theoryising

Missing Link
Frederick Rees Williams post office box 204 Pismo Beach Calif

All copyright broadcasting and otherwise reserved

LETTER FROM WILBERT C. CUNNINGHAM
TO
PROF. H. P. HUBBLE

CAT. # 0038

May 24, 1932

Prof. H. P. Hubble,
Mt. Wilson Observatory,
Pasadena, California.

Dear Professor:

Although man's persistence in spirit form is a much mooted question – quite incomprehensible, due to its invisibility and his lack of the true, natural basic law of life, nevertheless, the organic spirit of mankind truly continues after death. Truth is a power in itself. We act upon the theory that anything that is true can be made understandable.

Our material spiritual friends and helpers of the world external to man's negative sense of vision, (the great beyond close by) whose structural elements are directly relative to the crystalline chemical elements of the air, will become more tangible when the natural (electric) law accountable for this state of being, mode of passing on, perpetuity, etc., is made known to the people thru the work of the Spirit World.

The same physical (bi-electron) law is wholly responsible for mankind's origin on the Earth. To no other force or power whatsoever is man indebted for his crystalline germinal beginning on this planet.

I am not a college man, have had no training in the sciences; my knowledge of organic life, the world and universe could not have been gained from schools or text books. The thought contained in our work is transcendent, far in advance of the age. It shows the inspiration of a higher intelligence. Only one in direct communication with the Spirit World could have written the article enclosed herewith.

Money is needed to carry on the work of the Spirit World. We are in needful circumstances due to certain investments which, altho safe, owing to the general business depression are slow in materializing. We have stock in a mine company with valuable holdings in the Alma district, which we offer as security for a cash loan, or will sell all (1800 shares), or any part of same considerably under the company price.

This mine property (lead, silver and gold), near Alma, Colorado, adjoins the old Dolly Varden which made millions for its stockholders. Among other famous mines in this district are the well-known London properties, which according to a recent write-up in the Post, are great producers.

Hoping you will consider our proposition favorably, I am

Yours most sincerely,

Wilbert C. Cunningham
Direct Contact
Physico-Spiritual Writer Medium
1854 Stout Street
Denver, Colorado
Keystone 9595

LETTER FROM WILBERT C. CUNNINGHAM
TO
THE OBSERVERS AT MOUNT WILSON

CAT. # 0039

SECRETS OF THE SUN DISCLOSED BY THE SPIRIT WORLD

My information concerning the origin of the sun, the planets, the earth, and its organic life is derived wholly from the men and women spirits of the invisible world of mankind, who communicate that the primary germs of animal and vegetative life were formed in the processes of solar mundane crystalline elemental formation; that the planets, and the sun itself actually grew from the fundamental crystal structures naturally developed in the solar primary cubic crystallization system. In this system of natural creation the atom has no functional existence, practical use, or place in the sun, therefore the typical solar atom, which seems quite chaotic, is entirely neglected.

The sun's ultimate units are the positive and negative electromagnetic dualitites, or paired electrons (bielectron) which compose its orderly stratified crystalline elemental constitution, which substantial formation represents electricity in its natural bi-electron state the source of dynamo or commercial electricity being the kinetic flow of the sun-earth-moon inter-ray – or, specifically, intercellular crystalline light-ray – electronic circuit system. The fundamental fact of electricity is, natural electricity has a definite geometrical bi-electron constitution – thru cubical right angle triangle cross-ray vertical magnetic grip, solar electricity grips itself, so to speak, thus keeping within the cubic crystal confines, or relative systems of its own material force formation. From antecedent to consequent, electricity's a priori relativity were its own anterior or positive to negative changes. The solar and planetary bielectron units – formed of the uncreatable, indestructible, eternal force electricity – are not subvertible – fixed in the structural tension of elemental crystalline determination, are not subject to annihilation. Electricity's static cubical crystalline sun-formations are the exemplification of the electric unity of the stellar universes, and all they contain – of its own material-force restraint – power to hold the expansible properties of its violent electric nature within definite bounds. The sun-conformation is retained, and solar gravitational mechanics function thru the tensional properties of its triangular bielectron-ray concentric crystalline strata formations. Its bi-electronic hemispherical divisions conform to the opposite hemispheres of its primal bi-electron unit, the geometrical model of the primary formation, the plan of Old Sol's iron cubo octahedron center core. Since thru triangular bielectron light-ray crystalline formation, natural electricity retains the properties of its violent nature, necassarily the constituent entities, the bielectron-cell components, possessed a definite structure, were more than simple electric charges. The primitive antesolar electrons (more about them later) reached structural fixation at a certain moment in their everlasting geocentric movement toward the hub of all creation. The central Hub of All Creation turns in the depths remotely beyond the Milky Way and island universes. The stupendous proportions of its suns – in cubical quadrangular spacial distribution – are somewhat commensurable with the idea of time at no instant the beginning, of space in expanse no point an end. The work of the Spirit World will continue with new factors and features of the sun inside and out. Among the many features to be taken up, are the positive and negative differentiae of the solar electrons – the primitive antesolar electron's quadrilateral arrangement, and growth development from the inorganic cosmic stuff, or preuniversal antecrystalline electricity – factors and basic principles of natural electric law – the cause of solar and planetary directional and structural motion – structural and orbital ellipse – axial inclination and determination – formation and fixation of the solar and planetary elements – cubical crystalline strata dip and its relation to the angular distance of the moons and planets N. and S.

of the equators – application of Kepler's harmonic law to cubical crystalline structures – vertical descent of the "cosmic stuff", or rays, correlated with the perpendicularity of the structural universe – cause of the rise and fall of the oceanic tides – terrestrial and universal gravitation – bi-electron unification, the process, general significance in respect to the opposite hemispheres and parts of things, the universal biologic duality, the voluntary and involuntary male and femaale principles of nature, etc.

Respectfully,

Wilbert C. Cunningham
Intermediary.
May 15, 1932

Raleigh Hotel
1854 Stout St.
Denver, Colorado.

Boscobel Wis. 7/13. 1931.
To Mt. Wilson Observatory
Mt Wilson Calif.

Dear Sirs—
My first of this letter to you is to try and show you the Earth is not flat so that it turns round the Sun. You will please excuse me for saying so. I know what I am up against. Therefore you can see why I write to the people who are Scientists in Astronomy. Some one who is not satisfied with just Someone else ideas, But Some one that wants to go farther take it apart and see for themselves the Cause & why. I know that to write to just professors of any College or School would be a waste of time. Because their knowledge only runs as far as they been told, and never investigate for themselves. You know the Earth is not flat, But will tell you a [illegible] reasons why it is not. If the Earth was flat and the Sun & moon revolved in a circle overhead there never would be an eclipse of the moon, There would be no full moon and we know if the moon goes round the Sun must [illegible] so that it appears to once every 24 hours. The Earth

CAT. # 0040

LETTER FROM JOHN ROUNDS
TO
THE OBSERVERS AT MOUNT WILSON
CAT. # 0040

Boscobel Wis. 7/13 1931.
To Mt Wilson Observatory
Mt Wilson Calif.

Dear Sirs

My object of this letter to you is to try and show you the Earth is not flat or that it turns round the Sun. You will please excuse me for saying so I know what I am up against. Therefore you can see why I write to the people who are scientists on Astronomy. Some one who is not satisfied with just someone else ideas, but someone that wants to go farther take it apart and see for themselves the cause & why. I know that to write to just professors of any college or school would be a waste of time. Because their knowledge only runs as far as they been told and never investigate for themselves. You know the Earth is not flat, but will tell you a few reasons why its not. If the Earth was flat and the Sun & Moon revolved in a circle overhead there never would be an eclipse of the Moon, there would be no full moon and we know if the moon goes round the Sun must to. Or that it appears to once every 24 hours. The Earth would never be (—) the Moon & Sun and that we can go North for enough to see the Sun appear to stand still. That every spot on our Earth is the high spot. by the level. There to many objections to that. Still this is being taught in the schools at Zion City S Cl. Now its also being taught the Earth turns round the Sun. Earth is North of the Sun and turns around the Sun at one position to the Sun. That the Polar Star turns in a circle East & West. with the Earth. That in surveying there is a variation of direction from one point to the other and also another variation at different times of the year astronomy also teaches there are two poles. 6% degrees apart. Now don't get scared you won't have to change the map of the World because its made by the sunlight. All you change is the way in figuring this change. now for your consideration of the Earth is round how do you figure the Earth is tipped. Tipped from what. means just this Our Earth is called tipped from the Polar Star towards the Sun or plainly speaking tipped or another Pole an imaginary pole 6% degrees from the Earth and to aggree with the Sun or Our Earth turning round the Sun. But if the Earth turns on its axis there can be but one pole. The Earths axis any other any other place away from that center must revolve be any where in a circle 6% degrees away. Now for the sake of a shorter letter and explanation I leave out the North Pole entirely and go by the Worlds axis. Our Earths magnet is not the Sun but that Polar Star. It holds one face to that star night & day the year round. Our Sun turns round our Earth not exactly square or at right angles to our Earth & that Star but at an angle that would cause the difference in direction between the two poles. Our Earth holds one point to that Star and goes round it the same as our Moon does to the Earth. All our heavens & Earth revolve round that Star but the same as fixed stars so far as North & So are concerned. Our Sun turns round the Earth one way to the Earth makes no difference where the Earth is up or down or on the other side of that star. But the rest of the stars turn also. It makes no difference how long it takes for the Earth to travel round that star so far as our seasons are concerned. and it makes but little difference to the looks of it if the Earth turned round the Sun or the Sun round the Earth. But there is a difference that you can notice by the stars in line with the Sun & Earth. if the star in line or not. Our Earth can go round the Sun & still be in line. But not the sun round the Earth. Those stars will appear to be North or South of the Sub. and that star may possibly turn round the Sun but slowly. But as the Earth turns round that Pole Star those East or west stars appear to turn round our Earth. They are turning with the Earth round that Star. But when our Earth is 1/2 way round that star that was in the west will be in the East.

It will not be but our Earth turning on its axis make it appear so. There is no Suns or Earths turning on their axis North & South only one and it stands on end and does not affect any Star or Moon only its own. In Surveying its always that Pole Star Then a variation allowed for difference in direction and a variation at different times of the year. To correspond to the Earth turning round the Sun. Its against gravity to think the Sun is standing still. It turns on its axis opposite to the Earth or from East to west. It may be larger than the Earth but not so heavy there is no matter there. and it may be the Sun is not as large as our Earth & the atmosphere round the Sun burning. Our Sun shines a steady shine & the three motions it has at a terrible rate of speed may cause it to shine. its fine all O.K. but a different kind of fire than a fire caused by our match. Our fire. There is a Male & female to everything and our own sun & water are those two first. Now as long as I am at it I want to tell you my idea of a possibility that could happen. Not that I said it did but could of a possibility. I think it possible there was such a thing as a flood. That Noah was a scientist and astronomer or was told of these things or possibly of what might of happened before that at one time there was no water on our Earth. Not till the Moon came or we came to the Moon. That Noah knew some way there would be a flood. But not knowing what would happen when the Earth & Moon met. That all history before the flood was written of and on the Moon That that boat just fell off of the Moon to this Earth. It dosent seem possible a flood of that and could be any other way. and here arizes a question. can it be possible the missing link was those astronomers of the Moon. If that is so how can you blame the almighty. I would like to hear from you on this idea. I have written to the Observatory of Greenwich, England. Berlin Germany, & Athens Greece & you. and am going to send one to Poland, & France.

But I would like your opinion and thank you.

I am Yours Trully.
John Rounds.

It makes no difference where the Earths axis is that or the Earth turning round that Star. The Sun, belongs to this Earth and revolves round it the same as our Moon only opposite. The Sun is North more than South, Its because the Sun is more at right angles to the Earth & that Star in our winter is why its closer, Our Sun never shines over the South Magnetic Pole as far as it does over the North Magnetic Pole. If you draw a line through the Earth to a point as far South of the Equator and stand on that line at right angles your head will point in the same direction as the people on the Equator. Then your head will point South of the Sun, at this time of year, and if the Earth is North & stays North of the Earth only as it turns on its axis. The Sun causes the light & seasons and it all depends on how it turns round the Earth what angle if we have any seasons at all or not.

Letter from May Bernard Wiltse
to
Dr. George E. Hale

CAT #0041

Venice, California
4156 Neosho Avenue
December 3, 1932

Dr. George E. Hale
Mt. Wilson Observatory
Pasadena, California

Dear Doctor:

I just read in the paper that you have won the "Copley Medal" presented by the Royal Society of London. This Medal is a "Magnet" with a magnetic field.

In 1916 I went to Washington, D. C. and transmuted silver into gold for the United States government and I have their reports. BUT IT WAS HUSHED up for reasons I cannot explain.

At that time I was corresponding with one of your greatest astronomers in the United States—Doctor Ricard of Santa Clara College.

After returning from Washington I stopped writing to him, but one day I read in the Sunday paper a great article written by him about the sun being a magnetic Magnet and about the sun spots just exactly what I had written to him years before. I immediately sat down and asked him if he remembered what I wrote to him just what he had given to the public as his discovery and the wonderful scientist and man answered this question and here is a copy of his wonderful letter.

Copy

University of Santa Clara
Santa Clara
California

February 18, 1927

Dear Madam: I fully remember you but had forgotten the wonderful things you then said. From now on I shall be able to understand you better, although I feel my capacity is not up to your standard. I am glad to know you long ago discovered ALL the wonderful things that modern science is daily discovering.

You must be a wonderful woman.

Yours sincerely,

J.S. Ricard

This great man was big enough to acknowledge that I have discovered "ALL of the secrets of nature, not one, BUT ALL" and he KNEW what he was saying. And he did not feel hurt when I told him he had just told in the Newspaper just what I had written to him before.

Now Doctor Hale do you not remember me I wrote to you many years ago that I HAD DISCOVERED the sun was a magnet and the sun spots are 12 ducts and are filled with chemical like a cartridge in a gun and they are called Comets. Comets are NOT wanderers in space they are the feeders for the sun rays or Cosmic Rays the alchemists call them. I KNOW this is true because I have seen the comets go back into the sun many times with my naked eye. Any scientific man ought to know better than to say that an eclipse is caused by one body passing through the shadow of another. How any man with a thinking brain can say such a thing I CANNOT understand. Because Doctor Hale you know no shadow can CAUSE the wonderful eruption of the sun every time there is an eclipse of the sun IT MUST COME FROM THE CENTER OF THE SUN. As was well KNOW to the ancient astronomers and from them I have received ALL of my great knowledge NOT from any modern man. BUT from the books you sneer and scorn. THERE IS THE ONLY TRUE KNOWLEDGE

YOU WILL EVER KNOW. YOU NEVER WILL DISCOVER anything from looking through your telescope because you m must experiment in the laboratory HERE BELOW and KNOW ALL Electrical phenomena TO UNDERSTAND GOD'S Electrical machine.

I KNOW just exactly the working of God's Dynamo and I am going to write to the Royal Society of London and tell them exactly WHY the sun is a Dynamo and the Universe a Magnetic Field surrounding the sphere. Why you astronomers will not acknowledge my works I CANNOT understand. Are you jealous of me as a simple woman or what?

I have written to you so many times asking for an interview but you ignore my letters. Doctor Ricard did NOT, he answered every letter I wrote, and marveled at my work of the ancient alchemists and TRUE astronomers. I have written to many Scientific Journals and told them to keep my letters on file so as to KNOW I WAS FIRST in discovering the LAWS of nature before any modern scientific man did. So many of them know today that you are NOT the first, because I discovered the sun was a magenet in 1908 studying Electrical phenomena and NATURE.

Of course, I am an unknown woman without money or honor BUT I can fight and I SHALL until I am recognized by the colleges of the world.

I am writing to Oxford, Cambridge and France and Sweden to the College of Letters, in fact I was a candidate for the Prize but was too late in getting my works there.

I shall try next year. I sent you one of my books last year but did not hear from you whether you received it or not.

I would like to visit you or have you visit me and let me explain about my work.

Doctor Hale

Yours for TRUE science

Mrs. May Barnard Wiltse

Studebaker Big Six

TONOPAH-MANHATTAN STAGE
CLARK JAMES, Owner

Tonopah, Nevada September The Io 1929

Mount Wilson Observatory

Pasadena California

Gentelmen

Would you Please give us the Name of the Grate Estronminer that Invented the Interfeer Metor and the Name of the Instrument that you turn the light into and it tells how fast the big suns are traveling

Thanking you for anney information

Wee remane yours Trulry

Clark James

CAT. # 0042

LETTER FROM CLARK JAMES
TO
THE OBSERVERS AT MOUNT WILSON

CAT. # 0042

September The Io 1929

Mount Wilson Observatory
Pasadena California

Gentlemen

Would you Please give us the Name of the Grate astrominer that Invented the Interfeer Metor and the Name of the Instrument that you turn the light into and it tells how fast the big suns are traveling

Thanking you for anney information.

Wee remane yours Trulry

Clark James

St John
March 26/1933

Dear Sir

Just a few lines
to let you know that
I am Interested in Astronomy
I have did quite a lot
of reading on it and I am
really interested in it. I
had quite a bit of
confidence in Materialism
I believe myself the
whole Universe is substance
But what I would really
like to know is will
Astronomy get a person
anywhere is there any use

CAT. # 0043

LETTER FROM ALBERT JIGGEY
TO
THE OBSERVERS AT MOUNT WILSON
CAT. # 0043

St John
March 26 1933

Dear Sir

Just a few lines to let you know that I am Interested in Astronomy I have did quite a lot of reading on it and I am really interested in it. I have quite a bit of confidence in Materialism I believe myself the whole Universe is substance But what I would really like to know is will Astronomy get a person anywhere is there any use in a person studying it. Will it put you in an unmentally condition But I know that the more you read up on it the more you get Interested I often wondered myself if there is life in the Milky way and those constellations (Leo) Nebula in Orion Would you please give me some kind of basis to the Knowledge of astronomy.

Yours Sincerely
Albert Jiggey

Mr. Albert Jiggey
200 Paradise Row
Saint John
New Brunswick
Canada

PLATES

COURTESY MOUNT WILSON OBSERVATORY

PLATE II

George Ellery Hale

COURTESY MOUNT WILSON OBSERVATORY

Edison Petit

COURTESY OF MOUNT WILSON OBSERVATORY

PLATE IV

Seth B. Nicholson

COURTESY MOUNT WILSON OBSERVATORY

PLATE V

Joseph Hickox

COURTESY MOUNT WILSON OBSERVATORY

PLATE VI

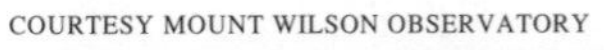

Larry Webster

COURTESY MOUNT WILSON OBSERVATORY

PLATE VII

The 100-inch Hooker Telescope

COURTESY MOUNT WILSON OBSERVATORY

PLATE VIII

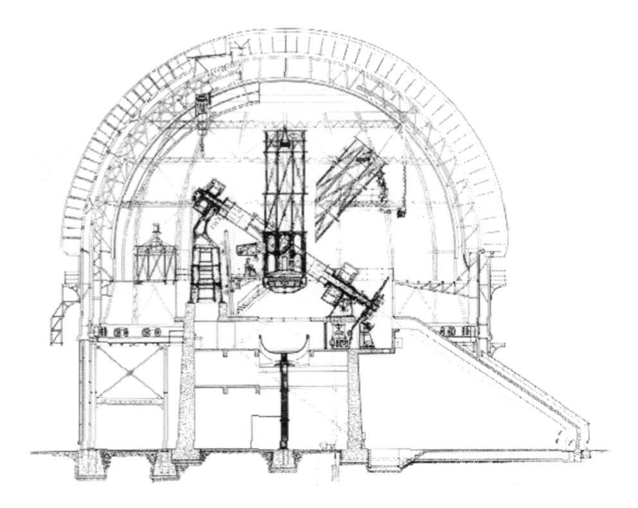

Diagrammatic drawing of the 100-inch Telescope

COURTESY MOUNT WILSON OBSERVATORY

PLATE IX

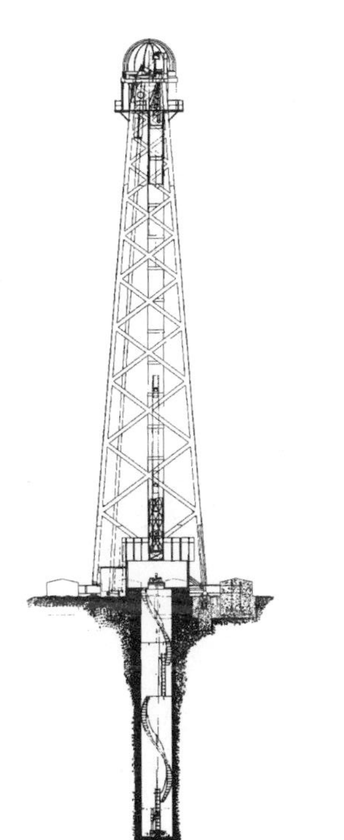

Diagrammatic drawing of the 150-foot Tower

COURTESY MOUNT WILSON OBSERVATORY

PLATE X

The 150-foot Solar Telescope

COURTESY MOUNT WILSON OBSERVATORY

PLATE XI

The 100-inch Telescope Under Construction

COURTESY MOUNT WILSON OBSERVATORY

PLATE XII

The 100-inch Telescope Under Construction

COURTESY MOUNT WILSON OBSERVATORY

PLATE XIII

The 100-inch Under construction

COURTESY MOUNT WILSON OBSERVATORY

PLATE XIV

The 100-inch Hooker Telescope

COURTESY MOUNT WILSON OBSERVATORY

PLATE XV

Haley's Comet

COURTESY MOUNT WILSON OBSERVATORY

PLATE XVI

The Snow Solar Telescope

COURTESY MOUNT WILSON OBSERVATORY

PLATE XVII

The 60-inch Reflector

COURTESY MOUNT WILSON OBSERVATORY

PLATE XVIII

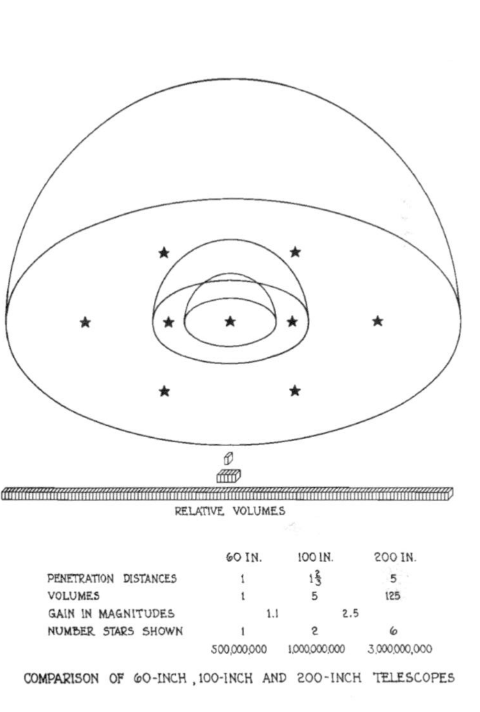

COMPARISON OF 60-INCH, 100-INCH AND 200-INCH TELESCOPES